Springer Aerospace Technology

The series explores the technology and the science related to the aircraft and spacecraft including concept, design, assembly, control and maintenance. The topics cover aircraft, missiles, space vehicles, aircraft engines and propulsion units. The volumes of the series present the fundamentals, the applications and the advances in all the fields related to aerospace engineering, including:

- structural analysis,
- aerodynamics,
- aeroelasticity,
- aeroacoustics,
- flight mechanics and dynamics
- orbital maneuvers,
- avionics,
- systems design,
- materials technology,
- launch technology,
- payload and satellite technology,
- space industry, medicine and biology.

The series' scope includes monographs, professional books, advanced textbooks, as well as selected contributions from specialized conferences and workshops.

The volumes of the series are single-blind peer-reviewed.

To submit a proposal or request further information, please contact:

Mr. Pierpaolo Riva at pierpaolo.riva@springer.com (Europe and Americas)
Mr. Mengchu Huang at mengchu.huang@springer.com (China)

The series is indexed in Scopus and Compendex

More information about this series at http://www.springer.com/series/8613

Min Zhao

Prediction and Validation Technologies of Aerodynamic Force and Heat for Hypersonic Vehicle Design

国防工業出版社
National Defense Industry Press

Springer

Min Zhao
China Academy of Launch Vehicle Technology
Beijing, China

ISSN 1869-1730 ISSN 1869-1749 (electronic)
Springer Aerospace Technology
ISBN 978-981-33-6525-4 ISBN 978-981-33-6526-1 (eBook)
https://doi.org/10.1007/978-981-33-6526-1

Jointly published with National Defense Industry Press
The print edition is not for sale in China (Mainland). Customers from China (Mainland) please order the print book from: National Defense Industry Press.
ISBN of the Co-Publisher's edition: 978-7-118-11739-4

This Springer imprint is published by the registered company Springer Nature Singapore Pte Ltd.
The registered company address is: 152 Beach Road, #21-01/04 Gateway East, Singapore 189721, Singapore

Foreword

Aerospace is a high-tech industry and field that demonstrates national strength and carries national willpower. All major aerospace countries are actively exploring and developing new types of spacecraft as spacecraft is an important symbol of the development of aerospace, and the research and development of new types of spacecraft are aimed to solve many new scientific and technological problems. Spacecrafts in supersonic speed and hypersonic speed are currently the focus of attention and research of the major space powers around the world. An important issue for this type of spacecraft is the heat barrier or new heat barrier.

The problems of heat barrier are actually issue on aerodynamic heat, which is one of the important issues in aircraft research and design. Researchers and designers seek analysis, experimentation, and prediction on aerodynamic heat based on the recognition of aerodynamic heat mechanism, which requires new theoretical predictions, experimental simulations, and research flight tests developed so that the three can complement each other and perfect each other. There is a very complicated aerodynamic force and heat problem compared with hypersonic aircraft. The understanding of force and heat environment, force and heat response, experimental simulation, etc. is not clear, and the solution needs to be explored and innovated.

The author of this book has conducted years of research and proposed the theory of hypersonic aircraft aerodynamic environment, numerical simulation prediction and experimental verification technology, which is commendable. This book is not a general all-encompassing aerodynamic textbook, which carries out a comprehensive analysis on the aerodynamic heat environment of hypersonic flight, and puts forward some new predictions and experimental theories and technologies on this basis. It systematically summarizes and analyzes in this book the development of aerodynamic force and heat technology for hypersonic vehicles, the consistency between ground and flying, the overflow and drag reduction, the method of force and heat prediction, transition and turbulence, and advanced wind tunnel tests.

It analyzes in this book the related research on the aerodynamic heat technology and the consistency of the flying and the ground for the hypersonic aircraft and sorts out the development context. It describes the prediction method based on functionals in the aspect of prediction technology. It is hoped that functional algorithms can be adopted to analyze data recognition laws, replace the human brain to recognize massive data in the future, and introduce advances in the field of mathematics into engineering applications in time. This is also a path for the development of advanced aerodynamics research to artificial intelligence, which is forward-looking.

Test technology has long been in the position of scheme verification and performance test in engineering design. The engineering design has a decisive say and it has gradually fallen behind, although the status is very high. We need to understand the flow from knowing what it is to knowing why it is with the increasing pursuit of fineness and accuracy in engineering design. We dare not dig the margin to the end, dare not use the method of leapfrog innovation if only knowing the total performance, without knowing the flow details and the physical mechanism behind these performances. It describes in the book the advanced non-contact measurement technology of the wind tunnel in terms of experimental verification technology, which measures the aerodynamic force and heat distribution without destroying the flow itself, captures rich flow details, and provides a reliable basis for analyzing the flow mechanism and evaluating the simulation model, which provides a method for the test to move from knowing what it is to knowing why it is.

Science, technology, and engineering are mutually reinforcing and promoting relationships. There will be no direction for scientific research without engineering practice. There will be no breakthroughs in key technologies and the driving force for engineering development without scientific research. These prediction, and experiment techniques in the book have been born in the field of science and have gradually formed the prototype of engineering technology in the laboratory, but they have to pass through the arduous promotion process before they can be applied in engineering design. It is gratifying that the editor selected these new technologies with important engineering application prospects based on their long-term experience and technology accumulated in engineering practice, and committed to pushing them to engineering applications, so as to make beneficial exploration for integrating scientific research results into engineering practice.

A journey of one thousand miles begins with one step. This book may illuminate a short way for the development of new superb aircraft engineering technology. It is believed that this book is an important reference book for scientific and technological workers engaged in superb aircraft research and graduate students engaged in this field.

Beijing, China
January 2018

Du Shanyi

Preface

I have always been uneasy when launching this book. Aerodynamic force and heat environment is an endless topic. New prediction and validation methods include all the frontiers of technology. It is inevitable to brag if saying such a big topic has been grasped. However, I have always been holding a desire to express my accumulated experience and feelings over the years, and invite experts and scholars to discuss on the understanding of aerodynamic force and heat environment and engineering technology, as well as put forward our ideas on the future technological development.

The aerodynamic force and heat environment, which is a problem that must be solved in the first step of aircraft design, and it is also the performance that must be evaluated at the first moment of the aircraft flying into the space. Such an important field is facing an awkward situation. Aerodynamic force and heat that is the frontier of scientific research. Many scholars have devoted their entire lives to theoretical research. There are many talents having put forward theoretical models, established simulation methods, and improved experimental techniques in universities and laboratories. The vast literature is proposing new ideas every day. However, the reality we are encountering is that the engineering aerodynamic heat design and validation technology is far behind the frontier development of science. The aerodynamic heat environment prediction and validation technology are actually slow in development due to the persistence of the industry and the difficulty of transforming scientific ideas to engineering technology. It should be said that the transfer from scientific ideas to engineering technology is the bottleneck of the development of many industries, and the research on aerodynamic force and heat environment is no exception.

It does not systematically describe in this book about the aerodynamic force and heat, since there are already many excellent textbooks in this field. Instead, it focuses on and analyzes the application of new prediction and validation technologies in the aerodynamic heat environment engineering design and introduces several technical development directions which are mature in theory and have potential driving force in engineering application. It is hoped that readers are people with high aspirations and determination to change traditional techniques. They are interested in further in-depth study after reading this book and will nurture new technologies and new ideas with application prospects in the design process of injection again in engineering applications, who have the ability to transfer scientific theory into technological innovation.

It first analyzes the technological development context and discusses the development prospects in this book. The connotation and engineering background of more new technologies are introduced, and the advantages and disadvantages of the new technologies are explained from the perspective of application. Relevant content includes Chap. 1 Development of Hypersonic Aerothermodynamic Technologies, Chap. 2 An Overview of Ground-Flying Calibration Technologies for Hypersonic Vehicle, Chap. 3 Flow over Blunt Body and Drag Reduction Control.

A new prediction technology based on intelligent algorithms and numerical simulation is proposed in terms of new prediction technology. It represents the development direction of the future prediction technology although the current prediction technology does not have good universality and reliability. Related content includes Chap. 4 A Functional Based Prediction Method for Hypersonic Aerodynamic Force and Heat, Chap. 5 Numerical Method for Flow Transition Prediction.

It is likely to become an important new validation method in engineering design although the new test technology proposed in the book has not yet been popularized in engineering applications. The related contents include Chap. 6 Embedded Aerosphere Sensing System, Chap. 7 Advanced Wind Tunnel Measurement Technology.

These contents are far from covering the breadth of aerodynamic force and heat environment research. They are just a far-reaching beckon to the future development with limited contents, and a small step is taken to shorten the gap between technological innovation and scientific frontiers with limited cognition.

The future has been coming but it is not yet popular. I hope, actual great technicians can realize our common dream in this great era.

Editor
Min Zhao

Beijing, China
December 2017

Informative Abstract

This book is aimed at analysis on the development context of new prediction and validation technologies in the field of aerodynamic heat research and promotion of the application of new technologies.

The development of aerodynamic force and heat technology and the research of ground-flying consistency of hypersonic aircraft were analyzed, as well as the new progressions in the field of bluff body overflow control and drag reduction technology. A prediction method based on functionals was proposed, invariant laws in multi-dimensional data is sought using intelligent algorithms and in-depth discussion was carried out on a difficult and important direction, turbulence transition numerical prediction. The embedded atmospheric data sensing system is introduced in the test verification technology, through which it is expected to provide important measurement technology for flight tests, and it described the phosphor thermography technology for measuring surface heat distribution and the particle image velocimetry technology for measuring the whole field rate distribution are adopted in the technology of non-contact measurement of wind tunnel.

This book is a reference mainly for professionals interested in aerodynamic force and heat, aircraft design, experimental testing, fluid mechanics and the fields of aerodynamic force and heat.

Contents

Chapter 1
Development of Hypersonic Aerothermodynamic Technologies

1.1 Prediction of Aerodynamic Force for Hypersonic Flow

The aerodynamic characteristics of a hypersonic aircraft determine its trajectory and performance of flight, and the aerodynamic characteristics are impacted by the contour of the hypersonic aircraft. The aerodynamic characteristics of aircraft with different contours must be accurately predicted in the initial design stage of hypersonic aircraft. The main methods of researches on the aerodynamics of hypersonic aircraft, include 3 aspects of theoretical engineering calculations, numerical simulations and experimental [1].

1.1.1 Theoretical Engineering Calculation Hypersonic Aerodynamic Force

It is impossible to invest too much resources in numerical simulation and experimental research in the early stage of the development of hypersonic aircraft, therefore it is essential to develop rapid and effective engineering calculation methods for aerodynamic calculations.

The air viscosity mainly acts in the boundary layer much thinner than the shock layer for hypersonic aircraft, while the viscosity effect can be ignored in the shock layer. Therefore, most of the aerodynamic characteristics can be predicted with the analysis method of inviscid flow, especially the predication for lift force and pitch moments can obtain satisfactory results. However, we still need to consider the impact from viscous resistance if we want to obtain aerodynamic information more accurately. The aerodynamic engineering calculation is introduced as the following from the calculation method of non-viscous aerodynamic force and viscous resistance.

M. Zhao, *Prediction and Validation Technologies of Aerodynamic Force and Heat for Hypersonic Vehicle Design*, Springer Aerospace Technology,
https://doi.org/10.1007/978-981-33-6526-1_1

1.1.1.1 Non-viscous Aerodynamic Force Calculation

The calculation of non-viscous aerodynamic force includes three common methods of Newton fluid model method, tangent-wedge method/tangent-cone method and Dahlem-Buck method.

(1) Newtonian fluid model method

Taking the motion of fluid mass points as particles in linear motion, the normal momentum is completely lost and the tangential momentum is retained after the fluid mass points hit the surface of the object. So let's assume that, the pressure coefficient of the fluid acting on the surface of the object is only related to the included angle θ between the object surface and the direction of the incoming flow. There is the following expression for the pressure coefficient:

$$C_p = \frac{p - p_\infty}{\rho_\infty V_\infty^2/2} = 2\sin^2\theta \tag{1.1}$$

In this equation: p_∞, V_∞, ρ_∞ is respectively the inflow pressure, velocity and density, p is the surface pressure of the object.

The original equation of Newton pressure coefficient is suitable for the situation where Ma approaches infinity. Lees [2] considers the modification of the Newton equation by the finite Mach number, and proposes the correction equation as the following:

$(C_p = C\sin^2\theta_{p\max})$

$$C\frac{2}{\gamma Ma_\infty^2}\left\{\left[\frac{(\gamma+1)^2 Ma_\infty^2}{4\gamma Ma_\infty^2 - 2(\gamma-1)}\right]^{\frac{\gamma}{\gamma-1}}\left[\frac{1-\gamma+2\gamma Ma_\infty^2}{\gamma+1}\right]-1\right\}_{p\max} \tag{1.2}$$

In this equation: $C_{p\max}$ is the pressure coefficient of stagnation points; γ is the ratio of specific heats.

Newton equation and Lees-modified Newton equation carry out a comparison of the pressure coefficient distribution around the surface of the ellipsoid with the accurate time advance solution [3] when the incoming Mach number Ma = 8, which shows that the Lees correction equation matches the exact solution better with the flowing problem over blunt body contour in not too high Mach number.

(2) Tangent-wedge method/Tangent-cone method

Tangent-wedge method/Tangent-cone method [4] is used to predict the surface pressure distribution of two-dimensional plane/axisymmetric body in hypersonic flow field. The basic idea of the tangent-wedge method/tangent-cone method is that the pressure at any point on the object surface is equal to the pressure after the plane oblique shock wave/conical shock wave with the local inclination angle θi as the half wedge angle/half cone angle, which is so obtained from oblique shock wave relation/cone shock wave relation.

It should be noted for the tangent-wedge method/tangent-cone method that the local inclination of the object surface cannot be greater than the critical separation angle corresponding to the incoming Mach number. Therefore, the tangent-wedge method/tangent-cone method is only applicable to the contour of a pointed body with a contact shock wave.

The tangent wedge method solves the shockwave angle β at each panel according to the θi-β-Ma∞ relation of oblique shock waves.

$$\tan\theta_i = \frac{2\cot\beta\left(Ma_\infty^2\sin\beta - 1\right)}{Ma_\infty^2(\gamma + \cos 2\beta) + 2} \tag{1.3}$$

The shock wave pressure P and the object surface pressure coefficient C_p are obtained with the parameter relationship before and after the shock wave after solving the equation to obtain the shock angle.

(3) Dahlem-Buck method

Dahlem-Buck method is a combination of the Newton fluid model method and tangent-cone method. The modified Newton method is used when the impact angle is large, and the approximate method of tangent cone is used when the angle of impact is small.

$$C_p = C_{pD} \cdot \frac{c_{p,cone}(Ma_\infty)}{c_{p,cone}(Ma_\infty = 20)} \tag{1.4}$$

In this equation: C_p, cone is the pressure coefficient of the conical flow at the corresponding Mach number, as a correction for the introduction of the Mach number; Newton method or cone cutting method is selected according to different impact angles for C_{pD}.

Among them:

$$C_{pD} = \begin{cases} \left[\frac{1}{\sin^{3/4}(4\delta)} + 1\right]\sin^2\delta \\ C\sin^2\delta_{p\max} \end{cases} \tag{1.5}$$

In this equation: δ is the impact angle, which is the deflection angle of the airflow.

1.1.1.2 Calculation Method of Viscous Resistance

The calculation of viscous resistance here mainly refers to the calculation of frictional resistance coefficient. The engineering calculation of frictional resistance is to expand the surface of the aircraft into an equivalent flat plate and make use of the research results of the flat plate. The friction resistance of the aircraft is obtained by adding the friction resistance of the plate with the contour correction factor and the compressibility correction factor. The friction resistance of the flat plate at low speed is:

$$C_{xfp} = 1.328/\sqrt{R_{eL}} \tag{1.6}$$

In the equation: C_{xfp} is the double-sided coefficient of friction resistance of the flat plate.

The friction resistance of the hypersonic aircraft can be calculated after correction:

$$C_{xf} = C_{xfp} \cdot \left(1 + 0.03Ma^2\right)^{-1/3} \cdot \eta_\lambda \cdot S_f/S_{ref} \tag{1.7}$$

In this equation: Re_L is the Reynolds number calculated from the length of the aircraft; η_λ is the contour correction factor; S_f is the side area of the aircraft.

1.1.2 Numerical Simulation of Hypersonic Aerodynamic Force

CFD numerical simulation is an important means to obtain aerodynamic force of the aircraft in the design of hypersonic aircraft. Most of the existing CFD calculation methods can be used to work out satisfactory results in the calculation of aerodynamic force, and can better describe the dynamic characteristics of the flow field for the numerical simulation of the flow field of hypersonic over flow. However, it will affect the numerical calculation of aerodynamic force, due to the complexity of hypersonic problems, such as transition in high-temperature boundary layer, chemical non-equilibrium flow effect and high-altitude thin flow effect, etc., so such problems need to be further studied.

1.1.2.1 Effect of Reseau/Turbulence Mode

The influence of the reseau mode on the calculation must be considered, and effective transition mode and turbulence mode must be used due to the complicated wall flow of hypersonic aircrafts, in order to accurately calculate the aerodynamic force.

Alter [5] evaluated the quality of the structural reseau in hypersonic calculations and the effect of the reseau quality on the calculation results, believed that the orthogonality of the reseau is important for the accurate calculation of the frictional resistance of the object surface under the hypersonic situation, and defined a reseau quality evaluation standard, that is, 1 means the best reseau, 0 means the worst. The grid quality is required to be between 0.6 and 0.8 in general hypersonic calculation. Nichols [6] carried out further discussion on the wall surface calculation problem in hypersonic flow, and proposed a boundary condition of turbulent wall surface function suitable for hypersonic flow, so that good wall surface pressure and friction resistance distribution can be obtained in the range of $y+ < 100$. Brown et al. [7]

verified the turbulence model of hypersonic flow, and compared with the experimental data with consideration of the change of pressure and frictional resistance on the object surface. The analysis showed that the k-ω SST and k-ω turbulence models were able to deliver more similar results to the experiment.

1.1.2.2 Chemical Non-equilibrium Flow Effect

The gas on the wall of the hypersonic aircraft will undergo a chemical reaction due to the high temperature effect. The CFD control equation at this time needs to consider different gas chemical components, chemical reaction models, etc., when the chemical reaction time scale is equivalent to the flow characteristic time scale, that is, in a chemically non-equilibrium flow state. Gupta et al. [8] classified the thermochemical state of the stagnation area of a typical ball head with a radius of 0.305 m.

Hassan et al. [9] carried out an aerodynamic analysis with consideration of the chemical non-equilibrium effect on the "Apollo" re-entry capsule, studied the effects of chemical reaction and vibration excitation on the lift-drag ratio and trim angle of attack with consideration of a 5-component chemical reaction model in calculation ($Ma = 27$), and compared with ideal gas and test results.

It is shown in the result that the coefficient of lift-drag ratio calculated with consideration of the chemical non-equilibrium effect is smaller than the result calculated with consideration of the ideal gas, and it is in good agreement with the experimental results. Papadopoulos et al. [10, 11] studied the characteristics of the pitching moment of the space shuttle at different angles of attack with a 5-component chemical non-equilibrium model. It is shown in the result that it will delay the reversal of the pitch moment coefficient with consideration of the chemical non-equilibrium effect. Bisceglia et al. [12] carried out a numerical simulation study with consideration of the chemical non-equilibrium effect, in which the wall pressure distribution and friction resistance coefficient of flight conditions with Ma = 9–26, altitude 40–80 km were compared, and it was found that the agreement of the pressure distribution and friction resistance coefficient is good as it is affected by the incoming Mach number, compared with the test and flight data.

1.1.2.3 Lean Gas Effect

The air in high-altitude atmosphere is thin, and the flow does not meet the assumption of continuity, so the free molecular flow effect needs to be considered. DSMC (Direct Simulation of Monte Carlo) establishes a gas molecule motion model from a microscopic perspective, which can be used to calculate the flow field of rarefied flow in high altitude. Bird [13] calculated the lift-drag ratio coefficient of the space shuttle with this method, and the results are in good agreement with the experiment. Wu Wei et al. [14] also took the impact from the dilute flow effect into consideration, numerically study the aerodynamic characteristics of the three-dimensional ball head with the DSMC method, and obtained the pressure distribution of the flow field around the ball head. It is shown in the result that the low-density effect of the thin flow has a great influence on the surface pressure coefficient distribution and drag coefficient, and the ball head drag coefficient increases significantly.

1.1.3 Experimental Study on Hypersonic Aerodynamic Force

Aerodynamic test research on hypersonic wind tunnel is of great significance for design and performance evaluation of hypersonic aircrafts. The working environment of hypersonic aircrafts is extremely special, accompanied by high temperature, high enthalpy, violent chemical reaction and other effects, coupled with its complicated structural contour, which has caused difficulties in the numerical calculation of aerodynamic performance. The hypersonic wind tunnel aerodynamic test is still the main method at present to study the prediction of the aerodynamic performance of hypersonic aircrafts and to solve key aerodynamic problems [15, 16].

1.1.3.1 Aerodynamic Force Measurement Technology

The ground equipment used to carry out various types of hypersonic aerodynamic force tests mainly include at present conventional hypersonic wind tunnels, pulse-type hypersonic wind tunnels (including shock wind tunnels/gun wind tunnels and hypersonic pulse combustion wind tunnels) and ballistic target equipment, etc.

Conventional hypersonic wind tunnel, usually $Ma < 10$, has high working pressure and density, and is in relatively long test time (from seconds to minutes). For pulse hypersonic wind tunnel, although the simulated Ma is higher, up to 20, and the working pressure and temperature (8,000 K) are higher, the test time is short (in the order of milliseconds). The ballistic target is a kind of aerodynamic ground test equipment that realizes the free flight of the aerodynamic test model in still gas. The aerodynamic parameters of the model are obtained, especially the dynamic aerodynamic coefficients such as dynamic derivatives and *Ma* is maximum up to 15.

Holland, Engelund et al. [17, 18] carried out summary on the wind tunnel aerodynamic test of the American Hyper X series test aircraft, made analysis on the hypersonic test, supersonic/transonic test, and obtained the law that the aerodynamic force coefficient changes with the angle of attack in different speed ranges, which has important reference value for the test flight of hypersonic aircraft.

(1) Wind tunnel balance technology

Wind tunnel balance is the core and key equipment for aerodynamic test research on hypersonic wind tunnel. The level of wind tunnel balance technology will directly affect the accuracy and reliability of results from wind tunnel aerodynamic test [19]. It can be divided into strain balance, piezoelectric balance and optical fiber balance according to the principle of balance measurement, the different sensitive components.

Strain balance is most widely used in hypersonic aerodynamic force tests, and the technology is the most mature. However, the force range is small due to the harsh environment of hypersonic force measurement and the small size of the model, which attaches high requirements on the hypersonic force balance. There had been various built-in "external microbalances" developed at home and abroad at present, including functions such as temperature correction and dynamic performance compensation, external strain resistance balances. A force test has been carried out on the blunt body model under $Ma = 94$ and the results agree well with the theory [20]. The design of this kind of balance makes full use of modern finite element technology, with a lot of simulation analysis performed, and the structural strength and impulse response are reasonably estimated, which makes the design of the balance more reasonable.

Piezoelectric balance makes use of the piezoelectric effect from piezoelectric ceramics or crystals for measurements on aerodynamic force. The piezoelectric balance is featured by high rigidity, fast response, simple structure, wide load range and strong model adaptability. Many scholars from abroad have designed and researched such balances, which have been widely used in shock tunnels [21–23]. It is shown in Marineau [22] test results that the piezoelectric balance is 350% higher in frequency response than the traditional strain balance and 400% in sensitivity with the same accuracy. The measurement of lift, resistance coefficient and moment of force is within 5% of CFD numerical simulation results. Jia et al. [23] designed and manufactured a new six-component piezoelectric balance, which has passed static calibration and dynamic testing. It is shown in the test results that the non-linear error and repeatability error of the balance are 0 06% and 0 19%, respectively, with good dynamic response and following performance. Lv et al. [24] designed piezoelectric balances with finite element software (ANSYS), analyzed their natural frequencies and modalities and other structural dynamics information, so as to ensure the reliability of the balance structure, and obtained the balance by changing the piezoelectric material parameters for the best performance.

Optical fiber balance was a new type of test balance developed to adapt to the aerodynamic test research under plasma and electromagnetic conditions [25, 26]. There are two main types of optical fiber strain balances at present. One is manufactured based on Fabry-Pérot interferometer. Arnold Engineering Development Center (AEDC) of the United States has carried out a plenty of work on the design and development of such balances. It shows that this fiber balance has higher accuracy and anti-interference ability compared with the traditional strain balance test.

China Aerodynamics Research and Development Center (CARDC) also independently developed a prototype of a two-component optical fiber balance principle (Fig. 1.1) based on this principle. It has been shown in the research that such balances are completely immune to electromagnetic interference and are not sensitive to temperature changes, so they are of high sensitivity, corrosion resistant, electromagnetic interference resistant, which can work normally in high temperature environment. The other is manufactured based on fiber grating, in which the strain is induced by the drift of the reflected wavelength of the fiber grating. Edwards [27]

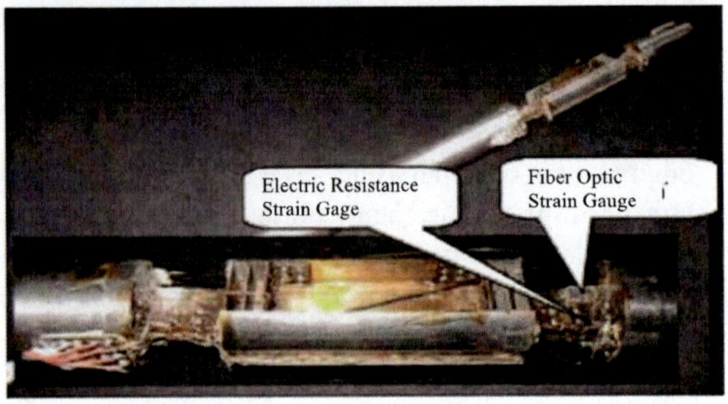

Fig. 1.1 Model machine made by CARDC in two-component fiber balance principle

added a fiber optic sensor to a six-component tail-bearing strain balance in order to facilitate comparison with traditional strain balances, and the normal force and pitch torque were measured for a blunt body. It is shown in the results that the optical fiber balance is with high accuracy and sensitivity, and can better adapt to high temperature conditions. Vasudevan [26] tested the friction resistance of the flat plate under hypersonic airflow with a fiber balance, and achieved good results.

Shenyang Aerospace University developed a five-component fiber grating force balance and carried out dynamic calibration experiments. It is shown in the results that the fiber grating balance has the same accuracy as the traditional strain balance, but the balance is greatly affected by temperature. Therefore, the first class of optical fiber balance, that is, the balance made based on the principle of Fabry–Pérot interference has more prospects and is more suitable for the hypersonic harsh environment test.

(2) Data acquisition technology

The data acquisition system of hypersonic wind tunnel mainly consists of bus type and instrument type. The bus type includes bus data acquisition board of PCI, ISA and bus platform data acquisition system of VXI, PXI.

It is foreseeable that the existing data acquisition system will continue to function during the next decade or so due to the periodicity of technological development and equipment updating. At the same time, it will be more commonly used in wind tunnels for the PXI bus data acquisition system, and some embedded field bus data acquisition systems will also be gradually applied in wind tunnels.

From the point of view of requirements on test accuracy index, the static measurement accuracy of the system should achieve 0.01–0.02%, and the dynamic measurement accuracy should not be less than 0.1–0.5%. However, for the requirements on acquisition speed of the data acquisition system, the static data acquisition system of 100,000 times per second has fully met the needs of wind tunnel testing, while the dynamic data acquisition system may be developed to more than 20 MHz.

1.1.3.2 Classification of Aerodynamic Force Test

The test of full model force measurement is the main form of hypersonic wind tunnel aerodynamic force test. The longitudinal and lateral aerodynamic characteristics of the model under different Mach numbers and attitude angles are obtained through this test, such as lift, resistance, side force, pitch moment, yaw moment and pressure center.

The test of model free flight is to quickly place the model into the test section flow field with the model delivery device after the wind tunnel flow field is established, and record the movement trajectory of the model. It is featured by no support rod interference, and is particularly suitable for measurement and research on resistance, bottom pressure, body interference and static and dynamic aerodynamic characteristics.

The test of jet stream interference characteristic is used to determine the interference of the engine jet (internal flow) on the outflow of the aircraft, do as to rationally lay out the structure of the aircraft to reduce resistance and improve flight stability. The reaction control system (RCS) of the space shuttle and the attitude control, autonomous flight and other characteristics of the missile are related to the jet interference effect.

The hinge torque test measures the torque of the aerodynamic force of each rudder surface of the model to the rudder angle, so as to work out the rudder surface control efficiency and provide a basis for designing suitable control devices. The hinge torque measurement usually adopts full model. The model is supported by the tail rod, and the force and moment of the airflow acting on the rudder surface are measured with a balance built into the model.

The dynamic stability (dynamic derivative) test is to measure the change rate derivative of the aerodynamic coefficient and aerodynamic moment coefficient of the aircraft to the dimensionless rotational angular velocity or attitude angle of the aircraft. Dynamic derivative is a critical aerodynamic parameter that affects the flight dynamics of the aircraft.

1.2 Research Trends of Hypersonic Aerodynamics Force and Outlook

The aerodynamic force of hypersonic aircraft is a key indicator in the development of hypersonic aircraft. As mentioned above, there has been considerable progress in the study of aerodynamics no matter in engineering calculations, numerical simulations or experimental research. However, the requirements for aerodynamic testing are getting higher and higher with the continuous development of hypersonic technology, and the following issues are still more prominent:

(1) The engineering calculation of hypersonic aerodynamic force is a quick aerodynamic force calculation method, while traditional calculations often ignore

frictional resistance (abbreviated as FR) and only perform non-viscous resistance calculation. The reason is that the wave resistance of low supersonic aircrafts (generally *Ma* is between 1 2 and 3 6) is large and the friction resistance is very small. The wave resistance is focused while the friction resistance is ignored in the calculation of aerodynamic engineering, which mainly adopts Newton flow modeling method for non-viscous dynamic calculation. The Reynolds number of hypersonic aircrafts ($Ma > 5$) becomes smaller, and their frictional resistance is often not negligible, which may even be greater than the wave resistance. Therefore, the frictional resistance effect must be considered in the aerodynamic engineering calculations of future hypersonic weapons in China.

(2) The wall surface flow of hypersonic aircrafts is complicated, and the numerical calculation of aerodynamics based on the N-S equation in the continuous flow area has been affected by the reseau division, selection of turbulence mode and transition mode, which greatly affects the accuracy of numerical calculation. It is still an important research direction for the development of hypersonic aircraft although a lot of work has been carried out.

(3) The continuous flow area is calculated based on the N-S equation for the numerical calculation of aerodynamics, while the continuous flow assumption is not met for the high-altitude free molecular flow area, and the control equation of N-S equation is no longer applicable, therefore a new control equation needs to be developed for numerical calculation.

The flow control equation of hypersonic dilute flow (free molecular flow) is Boltzmann equation. The traditional numerical solution method encountered great difficulties in practical application due to the highly nonlinear nature of the Boltzmann equation itself. The Direct Simulation Monte Carlo (DSMC) method based on statistical thoughts establishes and simulates the process of molecular collision, migration and energy exchange from a microscopic perspective, which is the most capable method at present to simulate 3D hypersonic rarefied gas flow under real conditions. It has achieved good results in the analysis on prediction of aerodynamics force and aerodynamic heat and has been successfully applied to the calculation of the lift-drag ratio of space shuttle from the United States, with consistent results with the flight test data.

(4) As the transitional flow region is between the free molecular flow and the continuous flow, its numerical calculations are extremely difficult to process. The numerical calculation with NS equation of continuous flow is not applicable, while the DSMC method for the thin flow is limited by the reseau, time step, and number of simulated molecules and other factors. The calculation efficiency in the transitional flow area is low and the calculation amount is large, so it is not applicable. Research on efficient and reliable numerical calculation methods is needed for transitional flow areas.

There are currently two widely used methods for the numerical calculation of the transitional flow area. ① CFD algorithm with slip boundary conditions. The condition of no slip on the wall surface is not met for the transitional flow area, and the boundary conditions of the wall surface with velocity slip and temperature slip can be adopted to modify the CFD method based on N-S equation. This method can solve the boundary condition problem to a certain extent, but it is not suitable when the air is thinner. ② The calculation is carried out with coupling of CFD and DSMC method, with full play of their respective advantages, expanding the use of the two algorithms, so as to make them suitable for the calculation of flow in the transition area. The research on this kind of coupling algorithm is the research hotspot at present on the numerical calculation of the transitional flow region, and most of the research is carried out with the target to improve the stability of the data exchange between the two algorithms. The coupling algorithm can guarantee a high calculation accuracy, and is suitable for the numerical calculation of the transitional flow region with a fast calculation speed.

(5) The gas on the wall surface of the hypersonic aircrafts will undergo chemical reactions due to the high temperature effect, which will affect the calculation of aerodynamic forces. It is an important research direction for how to carry out numerical calculations with consideration of chemical reactions.

The chemical reaction effect cannot be ignored at this time when the time scale of chemical reaction is equal to the time scale of flow characteristics, that is, in a chemically non-equilibrium flow state. The CFD control equation needs to take different gas chemical components, chemical reaction models, etc. into consideration and carry on supplemental corrections to ensure the accuracy of numerical solutions. There are hot research issues of the chemical reaction model, the selection of chemical reaction components.

(6) Test on hypersonic wind tunnel aerodynamic force is an indispensable important technical means to provide reliable data for the design and performance evaluation of hypersonic aircraft. The working conditions are relatively poor of the hypersonic wind tunnel force balance. The requirements on balance temperature control are high, and electromagnetic isolation may be required. It is extremely critical to develop corresponding balance force measurement technology with high-precision, high-sensitivity and anti-interference.

Optical fiber balance is an important technical solution to solve the above issues, which has higher accuracy and sensitivity, and can better adapt to high temperature and electromagnetic interference. In addition, it is an important developing direction of balance force measurement technology.

1.3 Prediction of Aerodynamic Heat for Hypersonic Flow

The high temperature thermal environment generated by pneumatic heating becomes more and more harsh as the design flight speed of hypersonic aircrafts is greatly increased. Taking a hypersonic aircraft with $Ma = 6$ as an example, the instantaneous heat flux density can exceed 12 MW/m^2 at the cone of the front radome, and the stagnation temperature can reach 1200 °C. The temperature in most areas such as the airframe, wings, and vertical tail is between 750 and 1450 °C, according to the temperature distribution of various parts of the Space Shuttle from the United States as it crosses the atmosphere. There will even be a local high temperature area near 1800 °C. The high temperature generated by severe aerodynamic heating during high-speed flight will significantly reduce the strength limit of the hypersonic aircraft material and the carrying capacity of the aircraft structure, result in thermal deformation of the structure, damage the aerodynamic contour of the components and impact the safety performance of the aircraft structure. It can be said that it has become the key to the success or failure of aircraft development for the accurate prediction of the aerodynamic heat of hypersonic aircraft and the thermal protection and thermal strength of materials and structures.

Hypersonic aerodynamic heat prediction is a technical guarantee for the successful development of hypersonic aircrafts. It has brought difficulties to the accurate prediction of hypersonic aerodynamic heat due to the complex flight environment (air dissociation, flow transition, etc.), complex aerodynamic shape and other factors.

There is no a more general prediction method for hypersonic aerodynamic heat prediction in theory, and ground test equipment cannot completely simulate many factors. Hypersonic aerodynamic heat prediction methods can also be divided at present into engineering calculations, numerical simulations, and experimental research methods.

1.3.1 Engineering Calculation of Hypersonic Aerodynamic Heat

The engineering calculation method is based on the similar solution of the boundary layer equation, which is obtained through theoretical analysis and reasonable assumptions or through the semi-empirical method formed by theoretical analysis based on the experimental data. The engineering calculation method has been developed rapidly because of its simple calculation process and high calculation efficiency since it was proposed.

1.3.1.1 Axial Symmetry Simple Contour Aircraft with Zero Angle of Attack

(1) Calculation of heat flow of stagnation points

The boundary layer similar solution is used to obtain the surface friction resistance coefficient for the calculation of the stagnation heat flow of the aircraft. The Blasius surface friction resistance equation is adopted for the laminar flow. The Schultz-Grunow surface friction equation is used for the turbulent flow. The impact of high-speed airflow is analyzed with the Eckert reference enthalpy method, and the surface heat flux density is calculated with Reynolds analog relationship. There are specific application forms include the widely-used Fay-Riddell equation, and the Kem p-Riddell equation with consideration of the shock wave effect [28].

(2) Calculation of heat flow of non-stagnation points

There is no a similar solution for non-stationary point heat flow calculations, since the fluid parameters and thermodynamic parameters change along the object surface direction, therefore it is necessary to introduce the assumption that the rate of change of each parameter along the object surface direction is much smaller than the rate of change along the object surface normal line. Now the velocity profile and the normal coordinate can be transformed to some contents to make the velocity profile the same in the flow direction, that is, there is a locally similar solution, so that the calculation can be performed. Lees [29] the equation for the laminar heat flux density distribution of bluff bodies can be obtained according to the non-stationary local similarity solution and can be used to predict the heat flux density distribution of the ball head, cone and other parts. The results are in good agreement with the ground test data.

Vaglio [30] and Zoby et al. [31] also developed calculation equations of turbulent heat flux for different non-stationary points.

1.3.1.2 Complex Contour Aircraft with Angle of Attack and Wing/Rudder

There are many types of aircraft at present, which can be divided into different types according to different ways of classification. So far as aerodynamic heat studies are concerned, the contour of an aircraft can be divided into small width blunt ratio aircraft, large width blunt ratio aircraft, etc. according to the aspect ratio (Aspect Ratio refers to the Ratio of the width to the thickness of the aircraft) of its body. There are different aerodynamic heat calculation methods for these two complex aircraft contours.

(1) Calculation of aircraft in small aspect ratio

There is mainly the axial symmetry analogy method, the equivalent cone method, etc. for the engineering calculation methods for the large area of the aircraft body in small aspect ratio.

The axial symmetry analogy method is based on the assumption of small transverse flow, taking the streamline as a coordinate axis, and to simplify the three-dimensional boundary layer equation into a quasi-axial-symmetry boundary layer equation with the Manger transformation, so that each surface streamline corresponds to a different axis-symmetric, and each streamline is independent, which can be used to calculate the surface heat flow of the aircraft with a small angle of attack and a small aspect ratio. Hamilton et al. [32] established a set of methods with this method for calculating laminar flow, transition and turbulent heating rates of any three-dimensional blunt body in hypersonic flow for ideal gas and air balance gas. The results are in good agreement with the experimental results.

The equivalent cone method is to replace the flow with an angle of attack with the flow with an equivalent cone of zero angle of attack. It is equivalent to the local inclination of the object face with angle of attack relative to the incoming flow.

(2) Calculation of aircraft in large aspect ratio

The strip theory can be used for contour of the aircraft in large aspect ratio, that is, the aircraft is cut out of several two-dimensional slices parallel to the center line of the aircraft, and assuming that there is no interference between these slices. It is to calculate the surface heat flow of these two-dimensional objects regarding these two-dimensional slices as two-dimensional blunt heads objects, and finally obtain the heat flow of the entire aircraft surface.

Adams et al. [33] calculated the surface pressure, surface heat flow and surface streamline flow of hypersonic inclined sweep cylinder, delta wing and other models with the strip theory, and it is in good agreement after comparison with the wind tunnel test data.

1.3.2 Numerical Simulation of Hypersonic Aerodynamic Heat

Hypersonic aircraft may generally experience three different flow regions of free molecular flow, thin transitional flow and continuous flow when flying in the atmosphere. There is a great influence from three different flow regions on numerical value calculation of hypersonic aerodynamic heat.

Knudsen number *Kn* is a dimensionless parameter reflecting the degree of gas thinning, $Kn = \lambda/L$, λ is the molecular mean free path, *L* can be taken as the overall characteristic length of the aircraft or the thickness of the boundary layer. The gas flow can be divided into a continuous flow zone, a transition zone and a free molecular

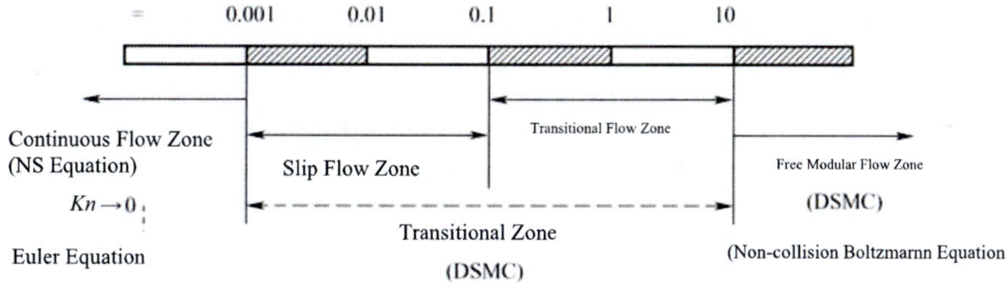

Fig. 1.2 Classification of degree of thin of gas

Fig. 1.3 Contour line distribution of Knudsen number with changes of altitude and flight speed

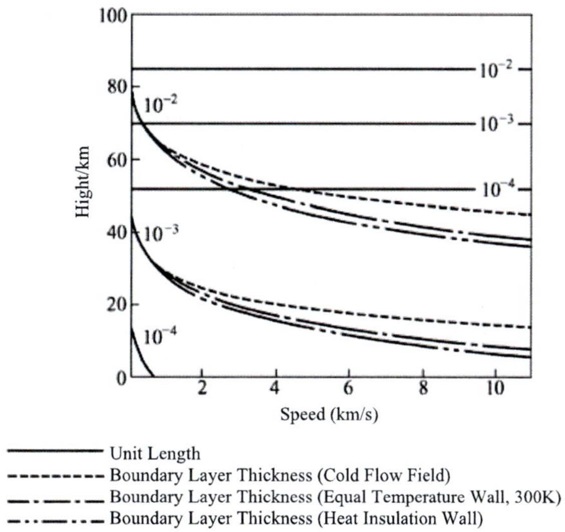

flow zone according to *Kn*, as shown in Fig. 1.2. 30–100 km is in the range of main cruising altitude of hypersonic aircrafts in the future in China. It can be seen from Fig. 1.3 that the flight area is mainly in the transition area, and the effect of rare gas needs to be considered.

1.3.2.1 Numerical Calculation of Aerodynamic Heat with Continuous Flow

It is to calculate the surface heat flux density of the aircraft by solving the N-S equation and its various simplified forms for the numerical simulation of the continuous flow area, including solving the viscous shock layer (VSL) equation, the parabolic N-S equation, the full N-S equation, etc.

There is great influence from numerical format, reseau form and turbulence mode on the solution of the equation, which in turn affects the prediction of aerodynamic heat.

Yan Chao et al. [34] studied the influence from different calculation formats (FDS format from Roe, FVS format from Van Leer, AUSM + format, and center difference format) on the accuracy of heat flow calculations, and believed that the AUSM + format is the most accurate for heat flow calculations. Pan Sha et al. [35] discussed the influence of reseau form on heat flow calculation. It was shown in the results that the heat flow calculation is extremely sensitive to the wall normal grid distribution due to the large wall temperature gradient and it was found that the convergence of aerodynamic heat calculation is much slower than the pressure and flow field convergence, that is, it is incorrect to take the equation residuals and pressure convergence as the basis for aerodynamic heat convergence. Zhang Xianghong et al. [36] carried out analysis on the influence of different turbulence models on the calculation of heat flux density, and it was found that the SST model and the k-ω model have good results in the calculation of heat flux density after the comparison with the experimental results.

1.3.2.2 Numerical Simulation of Thin Flow Region

The flow in the thin flow region no longer meets the continuity assumption. The governing equation of the fluid is Boltzmann equation. The traditional numerical calculation method is difficult to solve due to the high nonlinearity of the equation. Direct Simulation Monte Carlo method (DSMC) establishes gas molecule collisions, migration and energy exchange processes from a microscopic perspective based on statistical ideas. It is currently the most capable method to simulate the flow of a real three-dimensional hypersonic rare gas.

Moss et al. [37] calculated the hypersonic flow state of the double cone model and the hollow cylindrical skirt model with the DSMC method. It takes into account important phenomena such as shock wave boundary layer interference and boundary layer flow separation in the calculation. It was shown in the results that the calculation results are sensitive to reseau accuracy and physical model parameters, and the test results are in good agreement compared with the CUBRC test results. CARDC from China has also developed a thin gas dynamics calculation software based on the DSMC method, which has been verified by a large number of wind tunnel tests. It can be used for accurate prediction for all kinds of supersonic/hypersonic aircraft, plume, local aerodynamic interference and heat environment. The related research results are shown in Figs. 1.4 and 1.5

1.3.2.3 Numerical Simulation of Transitional Flow Area

It is extremely difficult for numerical simulation on the flow in the transitional flow region. The calculation of the N-S equation of continuous flow may fail locally or in

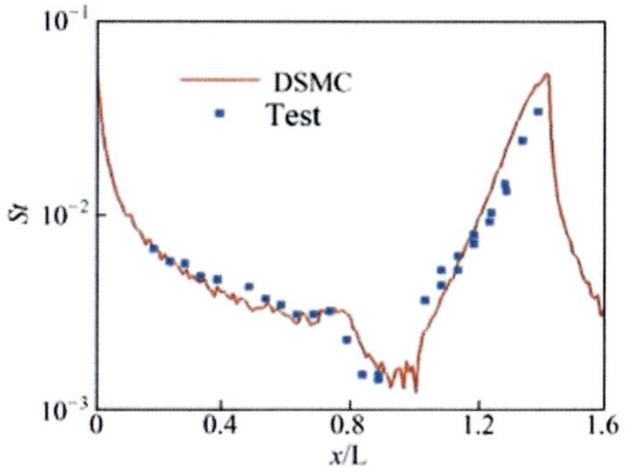

Fig. 1.4 Calculation and comparison experiment of heat flow on meridian

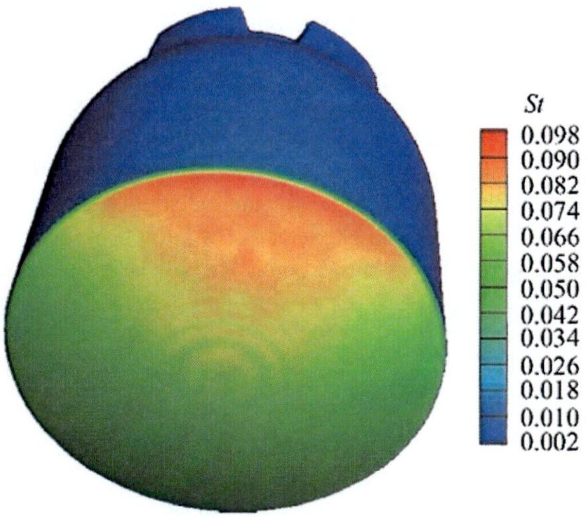

Fig. 1.5 Distribution of heat flow clouds on the surface of the re-entry capsule

all regions as the gas is thinner, resulting in larger errors. The calculation efficiency of the transitional flow area in DSMC method of rarefied flow is low and the calculation amount is large, so it is not applicable as it is limited by factors such as reseau, time step, and number of simulated molecules. There are three main treatment methods for this area at present.

(1) CFD algorithm for slip boundary conditions

It requires correction of the wall boundary conditions for the CFD numerical simulation based on the N-S equation in the transitional flow region. It needs to consider the effect of low density and thermal creep effect for the boundary conditions at this time, and the wall surface boundary conditions of velocity slip and temperature jump are adopted. The numerical simulation of N-S equation modified by the sliding boundary conditions can be applied to the calculation of the transitional flow region. The calculation result with consideration of the slip boundary is closer to the calculation result of DSMC.

(2) A unified cross-flow hypersonic algorithm based on Boltzmann equation

The effects of gas molecular density, collision frequency, and thermodynamic effects are taken into consideration based on the basic theory of Boltzmann equation. The discrete velocity coordinate method of gas kinetic theory is popularized and introduced according to the principle of microscopic molecular dynamics probability statistics, and an optimization method is introduced to determine discrete velocity coordinate points. The gas molecular velocity distribution function equation is transformed into a hyperbolic conservation equation can be numerically solved by the finite difference method of computational fluid dynamics (CFD). However, this method requires large amounts of calculation.

(3) CFD/DSMC coupling algorithm

The calculation is carried out with coupling of CFD and DSMC method, with full play of their respective advantages, expanding the use of the two algorithms, so as to make them suitable for the calculation of flow in the transition area. The data exchange method between the two algorithms is the key to this coupling algorithm. A variety of exchange boundary processing methods have been developed in order to suppress the statistical fluctuations generated during data exchange. Among them, it is widely used for the MPC coupling technology developed by Boyd [38]. This method can guarantee a higher calculation accuracy and a more rapid calculation speed.

1.3.3 Experimental Study on Hypersonic Aerodynamic Heat

The engineering calculations and numerical simulations on aerodynamic heat hypersonic aircrafts have been simplified to varying degrees on practical issues. It is necessary to conduct experimental research and provide data reference for aerodynamic heat calculation in order to obtain accurate aerodynamic heat data. There is great significance to the development of hypersonic aircrafts.

It mainly includes heating technology, temperature test, heat flow test, high temperature strain test, etc. for aerodynamic heat test technology. There are quartz lamp heaters, graphite heaters, and arc lamp heaters included in radiation heating devices for heat tests.

Graphite heaters can provide a higher temperature, but their heating rate is slow, which cannot meet the demands for high speed of surface heating of hypersonic aircraft. The arc heater can also provide a higher heat flux density, but the operation is unstable. Quartz lamps heating has high heating power, small size, small thermal inertia, and excellent electric control performance.

NASA Dryden from the United States simulated the pneumatic heating environment with a modular quartz lamp heating device, which can achieve a high temperature of 1,650 °C.

NASA Dryden also mainly uses thermocouples for the surface temperature test of the structure. As the main method for measuring the thermal environment under high temperature and high pressure, thermocouple has the characteristics of fast response, large output and stable performance. NASA Dryden developed the testing technology for blackbody light pipe temperature, which is adopted for accurate and repeatable temperature measurement of high-temperature composite materials, and completed the calibration of the blackbody light pipe temperature sensor in the blackbody calibration furnace. The error between the temperature sensor of the black body light pipe and the optical pyrometer is less than 1 °C in the temperature range of 800–1,500 °C. The black body light pipe has the advantages of electromagnetic radiation interference resistance, wide test range, high temperature resistance and so on [39].

Techniques on heat flow testing include point heat flow testing and large area heat flow testing. Electric heat flow sensors are mainly used for spot heat flow testing, with high accuracy, small size, high frequency response and other characteristics. Tao System and Virginia Tech jointly developed a new type of heat flow sensor in small size, high frequency response, which is used in high-speed spacecraft heat flow testing. NASA Glenn proposed to develop a double-sided, single-sided thin film Wheatstone bridge heat flow meter on an alumina substrate in 2002, which is easier to manufacture than a heat flow meter of thermopile type film and has a larger output signal.

Test for large-area heat flow includes infrared heat map, temperature-sensitive paint technology, thermochromic liquid crystal technology, etc. It can obtain the heat flow distribution of the entire model surface by optical measurement methods, which has the advantages of high spatial resolution and intuitive image, but its measurement and test accuracy are lower.

It mainly included foil strain gauges, welding strain gauges, wire wound strain gauges, and silica fiber strain gauges for the existing strain sensor types. It needs to research on the accurate and reliable strain test method under extreme ambient temperature for the heat test on hypersonic aircraft structure.

NASA Dryden from the United States has realized that this new test technology of optical fiber sensors has many advantages that traditional sensors do not have (high temperature resistance, electromagnetic interference prevention, etc.) over the recent years, which can meet the needs of hypersonic aircraft structural testing and began to fully develop high-temperature optical fiber testing technology.

1.4 Research Trends of Hypersonic Aerodynamics Heat and Outlook

(1) A combination is used of the engineering calculation method and numerical calculation method of aerodynamic thermal prediction

There are too many problems in the engineering calculation method of aerodynamic thermal prediction such as too many simplifications, difficult to calculate complex shapes, the calculation accuracy being limited. However, it requires a large calculation amount, calculation processing is cumbersome, and it takes too much time for the numerical simulation method. An effective solution is to combine both of the methods, numerically solve the value of non-sticky area outside the boundary layer, and calculate the theoretical equation in the viscous area inside the boundary layer, in order to obtain the heat flow distribution on the surface of the superb aircraft, which can save a lot of calculation time and is suitable for complex aircraft contour.

(2) Improve the heating capacity and test accuracy of ground test

There are two main difficulties in the test of aerodynamic heat prediction: simulation in high temperature environment and accurate testing of physical quantities in high temperature environment. It can basically achieve high temperature environments at present can basically achieve high temperature environments of quartz lamps, graphite heaters, arc lamp heaters and focused heating devices. The accuracy of the sensor still needs to be improved and it should have the capacity to resist electromagnetic interference (The air will be ionized in high temperature). The optical fiber testing technology can solve the above problems to a certain extent. The time and spatial resolution of the digital image need to be increased for large-area heat map testing, thereby improving the accuracy of the overall test.

(3) Development direction of heat protection technology for hypersonic aircraft in china

Most of the existing aircraft heat protection systems are simple in contour and short in protection time, which mainly adopt ablative thermal protection. However, it requires long-term heat insulation and heat protection when hypersonic aircrafts fly in the atmosphere for a long time. In addition, hypersonic aircrafts require good aerodynamic contour, and ablation protection will damage the aerodynamic contour, therefore, corrosion protection is not applicable for hypersonic aircrafts, and new heat protection methods need to be researched, such as considering the use of active heat insulation technology. There are sweat cooling, film cooling, and convection cooling for specific active heat insulation technologies. Ceramic composite materials and reinforced C/C composite materials should be considered as the main materials for ultra-high temperature insulation if heat-resistant materials are used, and metal insulation materials should be avoided.

The combination of active and passive insulation technology can be considered to combine the advantages of the two insulation technologies for the specific realization of the thermal protection of hypersonic aircraft.

The design trend of hypersonic aircraft heat protection system is "integrated design" at present, that is, the analysis on hypersonic aerodynamic heating environment, the high temperature fuselage surface radiating heat outward, the heat transfer of the fuselage structure inward, the material strength characteristics, the thermal response of the structure shall be combined, as well as the thermal expansion and deformation of the structure and the characteristics of the real environment of the internal cooling system of the structure. However, there are also many practical difficulties in this design, such as how to establish a theoretical and numerical model of multiple physical fields coupling under complex boundary conditions close to the real situation, and find a more accurate analytical solution method.

1.5 Summary and Outlook

The prediction on aerodynamic parameters under hypersonic flow has qualitative laws such as Mach number irrelevant principles, the prediction of pressure coefficients of Newton models, Lees correction, tangent cone method, etc. that have been established, the more complicated method of estimating the viscous effect due to hypersonic compressibility, boundary layer thickens and other characteristics, the calculation method of the friction resistance considering the influence of the boundary layer with the laminar flow and turbulent flow.

It mainly involves two aspects: overall aerodynamic heat and local aerodynamic heat for the research on aerodynamic heat. It becomes particularly important for the prediction of aerodynamic heat in the process of the development of hypersonic aircrafts at the same time, due to the complex shock wave structure in the case of hypersonic conditions, and the different interference modes of the shock wave and the boundary layer that will cause different degrees of local heat flow amplification.

What needs to be pointed out is that, the theory of blunt body under hypersonic flow is established with the research on the aerodynamic heating of the stagnation point surface layer and through the analysis of the influence of the pressure difference resistance coefficient and the friction resistance coefficient on the aerodynamic heating heat flow of the aircraft.

In addition, it is also faced with many difficulties and challenges for the numerical simulation and experimental research of hypersonic flow. An effective fluid medium model needs to be established according to the actual flight environment in terms of numerical simulation. Especially for the simulation of rarefied flow in the upper atmosphere, it is an inevitable problem to establish a highly accurate difference format of highly nonlinear control equations in a variety of complex chemical and thermal non-equilibrium situations a high-precision turbulence calculation model. And the accurate simulation of the heat flow on the surface of complex aircraft also needs to be established with a higher-precision calculation model. Therefore, it equips the numerical simulation of hypersonic flow with broad research prospects by chemical reaction effects, turbulent flow modes, thin gas effects, and intermittent shock wave solutions in the inviscid flow model, aerodynamic heating problems, etc. The focus of experimental research of hypersonic flow is the establishment of suitable wind tunnel devices, the development of parameter measurement equipment with high sensitivity and frequency response. The tough environment of hypersonic flow test, the inconvenient way to establish the super test environment (short wind tunnel test time, difficult to simulate high temperature thermal environment, etc.) has restricted the development of experimental research on hypersonic flow, although there are a variety of rapidly developing test equipment including optical fiber antennas.

There is close links between hypersonic flow field, aerospace and many other fields. Related research will continue to develop in the direction of the combination of experiment, theory and simulation in the future.

References

1. Anderson JD (2006) Hypersonic and high temperature gas dynamics. AIAA Educ 5:16–19
2. Lees L (1955) Hypersonic flow. In the fifth international aeronautical conference. Los Angeles, pp 241–276
3. Fritsche B (2001) Note on the application of SCARAB to the MIR re-entry. Proc Int Work-shop MIR Deorbit 5:92–101
4. Gentry AE, Smyth DN, Oliver WR (1973) The mark IV supersonic—hypersonic arbitratry body program, volume II, program formulation. Technical report, AFFDL-TR-73-159, pp 55–59
5. Alter SJ (2004) A structured - grid quality measure for simulated hypersonic flows. AIAA Paper 15:61
6. Nichols RH, Nelson CC (2004) Wall function boundary conditions including heat transfer and compressibility. AIAA J 42(6):1107–1114
7. Brown JL (2002) Turbulence model validation for hypersonic flows. In: 8 th AIAA/ASME joint termophysics and heat transfer conference, Saint Louis, MO
8. Gupta RN, Yos JM, Thompson RA et al (1990) A review of reaction rates and thermodynamic and transport properties for an 11-species air model for chemical and thermal nonequilibrium calculations to 30000 K. NASA STI/Recon Technical Report N 90:27064
9. Hassan B, Candler GV, Olynick DR (1992) The effect of thermo-chemical nonequilibrium on the aerodynamics of aerobraking vehicles. In: 27th AIAA thermophysics conference
10. Papadopoulos, Periklis E, et al. (2001) CFD, simulations in support of shuttle orbiter contingency abort aero-dynamic database enhancement. AIAA Paper 23:15–19
11. Prabhu DK, Papadopoulos PE, Davies CB et al (2003) Shuttle orbiter contingency abort aerodynamics, ii: realgas effects and high angles of attack. AIAA-Paper 16:1248
12. Bisceglia S, Ranuzzi G (2005) Real gas effects on a planetary re-entry capsule. In: AIAA/CIRA 13th international space planes and hypersonic systems technologies, AIAA Paper, vol 3385, p 7
13. Bird GA (1990) Application of the direct simulation Monte Carlo method to the full shuttle geometry. AIAA and ASME, joint thermophysics and heat transfer conference, 5th, Seattle, WA
14. Xiong Wu, Weifang Chen (2002) DSMC simulation research on the aerodynamic characteristics of the ball head in the transition area. Space Return Remote Sens 23(3):1–5
15. Guobiao C, Dajun X (2012) Hypersonic aircraft technology. Science Press, Beijing
16. Hanxin Zhang (2004) Hypersonic aerodynamic force test. National Defense Industry Press, Beijing

17. Holland SD, Woods WC, Engelund WC (2001) Hyper-X research vehicle experimental aerodynamics test program overview. J Spacecraft Rockets 38(6):828–835
18. Engelund WC, Holland SD, Cockrell CE et al (2001) Aerodynamic database development for the Hyper-X airframe-integrated scramjet propulsion experiments. J Spacecraft Rockets 38(6):803–810
19. Dexin He (2001) Wind tunnel balance. National Defense Industry Press, Beijing
20. Vadassery P, Joshi DD, Rolim TC et al (2013) Design and testing of an external drag balance for a hyper-sonic shock tunnel. Measurement 46(7):2110–2117
21. George RD, James FM (1968) An improved piezo-electric balance for aerodynamic force measurements. IEEE Trans Aerospace Electron Syst AES 4(3):351–359
22. Marineau EC (2011) Force measurements in hypervelocity flows with an acceleration compensated piezoelectric balance. J Spacecraft Rockets 48(4):697–700
23. Jia Z, Gao Y, Ren Z et al (2014) Research on the characteristics of a piezoelectric balance for the dynamic motion device. Int J Ind Syst Eng 18(3):364–381
24. Lv Z, Liu X, Zhang Y, et al (2015) Simulation and analysis of a piezoelectric balance. Electromagn Non-destruct Eval 40:141
25. Chandrakishore M, Srihari GK, Rudresh C, et al (2005) Aerodynamic load measurements at hypersonic speeds using internally mounted fiber-optic balance system. Instrumentation in aerospace simulation facilities, 2005. iciasf05 21st international congress on. IEEE, pp 119–122
26. Vasudevan B (2005) Measurement of skin friction at hypersonic speeds using fiber-optic sensors. In: AIAA CIRA 13th international space planes and hypersonics systems and technologies conference, p 3323
27. Edwards AT (2000) Comparison of strain gage and fiber optic sensors on a sting balance in a supersonic wind tunnel. Virginia Polytechnic Institute and State University, Virginia
28. Chenggong L, Hengzhi F, Qiao Y (2002) Aerospace materials. National Defense Industry Press, Beijing
29. Lees L (1956) Laminar heat transfer over blunt-nosed bodies at hypersonic flight speeds. J Jet Propulsion 26(4):259–269
30. Vaglio-Laurin R (2012) Turbulent heat transfer on blunt-nosed bodies in two-dimensional and general three- dimensional hypersonic flow. J Aerospace Sci
31. Zoby EV (1982) Comparisons of free-flight experimental and predicted heating rates for the Space Shuttle. AEDC Report
32. Dejarnette FR, Hamilton HH (1975) Aerodynamic heating on 3-d bodies including the effects of entropy-layer swallowing. J Spacecraft Rockets 12(1):5–12
33. Adams J, John C, Martindale WR (1973) Hypersonic lifting body windward surf-ace flow-field analysis for high angles of incidence. AEDC Report
34. Chao Y, Jianjun Y, Junzhe L (2006) Research on some issues of format and reseau effect in heat flow CFD calculation. J Aerodyn 24(1):125–130
35. Sha P, Dinghua F, Guohao D et al (2010) Reseau correlation and convergence in aerodynamic thermal simulation. J Aeronautics 31(3):493–499
36. Xianghong Z, Yizhao W, Jiangfeng W (2012) Numerical simulation research on aerodynamic characteristics of axisymmetric reentry module. J Appl Mech 29(3):284–290
37. Moss JN, Bird GA (2005) Direct simulation monte carlo simulations of hypersonic flows with shock interactions. AIAA J 43(12):2565–2573
38. Schwartzentruber TE, Scalabrin LC, Boyd ID (2007) A modular particle-continuum numerical method for hypersonic non-equilibrium gas flows. J Comput Phys 225(1):1159–1174
39. Edmund H, Everlyn C, Patricia P (1999) Research engineering annual report. NASA Report

Chapter 2
An Overview of Ground-Flying Calibration Technologies for Hypersonic Vehicle

2.1 Aerothermodynamics Environment of Hypersonic Vehicle

The conditions of flight environment for hypersonic aircrafts are all airspace and large velocity field. It is necessary to develop an environment prediction and verification method with high precision that comprehensively considers various physical effects under real flight conditions as ground design is difficult to fully simulate the high-altitude environment and flight conditions.

It is bound to be significantly affected by the aerodynamic force and heat environment during the process of launching from the ground to the reentry and the flight of a hypersonic aircraft at high speed in the atmosphere. The accurate prediction on the aerodynamic force and heat characteristics of the aircrafts can improve the overall performance and key indicators of the spacecraft from the source of design. To be more specific, the precise prediction on aerodynamic lift and drag is conducive to the accurate estimation of the range. The accurate prediction on aerodynamic torque is conducive to the refined design of attitude control and improves the maneuverability and strike accuracy. The accurate prediction on moment of pneumatic control surface hinge can avoid excessive margins in mechanism design, steering gear selection, and power supply selection. The prediction accuracy on aerodynamic loads directly affects the design results of load and structure. Too conservative structural design will bring unnecessary mass redundancy, which will affect key indicators of flight range, maneuverability, and throwing quality, etc. The accurate prediction on heat environment can effectively reduce the mass margin of the heat protection structure. Therefore, it requires the development of highly accurate prediction methods on aerodynamic force and heat environment to improve the performance of aerospace crafts.

The influence of the radiation environment is usually ignored when the hypersonic aircraft is flying outside the atmosphere, due to the short flight time. With the deepening of research on the radiation environmental effects of various spacecraft,

M. Zhao, *Prediction and Validation Technologies of Aerodynamic Force and Heat for Hypersonic Vehicle Design*, Springer Aerospace Technology,
https://doi.org/10.1007/978-981-33-6526-1_2

it was found that the superb vehicle will be affected by the radiation of high-energy particles even if the mission time is short and the orbit is low, especially the single event effects, which is a probabilistic event. A sensitive device may fail when it reaches its single particle effect threshold when it is hit by high-energy particles. In addition, the continuously decreasing size and increasing integration of electronic devices have risen the risk of single-particle effects. Therefore, there has been more and more attention paid to the effects of radiation environmental effects in various flying missions.

Electronic equipment may be damaged, and its functions are seriously affected during flying missions, especially when strong solar activity erupts. The technology of accurate prediction on flying radiation environment effect can be used to carry out analysis on the radiation effect of flying high-energy charged particles in the spacecraft flight environment, accurately assess the impact of high-energy particles on the electronic system, improve reliability, and support the leapfrogging of the spacecraft design level.

Compared to subsonic or supersonic flight, the high temperature generated by the flow blockage of the head of a hypersonic aircraft results in the vibration excitation, dissociation and even ionization of air molecules, which makes ordinary air a complex medium that continuously undergoes thermal chemical reaction. The microscopic changes of this kind of medium micell have an important influence on its macroscopic motion state by changing its chemical, physical and thermodynamic properties, and lead to the failure of conventional gas flow pattern, which makes the hypersonic flow phenomenon exceed what can be accurately predicted by the classic gas dynamics theory and has become one of the frontal subjects of modern gas dynamics research. The requirements on shape and structure, especially for advanced new types of aerospace aircrafts, are becoming more and more precise, and the aerodynamic layout and structural design are becoming more and more difficult. Therefore, the difficulty in predicting characteristics of the aerodynamic force/heat of hypersonic flow has become an important bottleneck with constraint on the research and development of aerospace aircrafts. The main difficulty recognized at present lies in the high enthalpy thermal chemical reaction gas flow show the flow characteristics of non-linear and non-equilibrium and multiscale compared with new prediction and validation techniques on aerodynamic heat environment of traditional subsonic, supersonic, hypersonic aircrafts.

So far, there is still a lack of mathematical and physical equations with proper accuracy, so wind tunnel test is still one of the main methods to obtain aerodynamic data of aircraft.

A large number of different types of hypersonic ground test equipment have been developed in the world, in order to develop hypersonic aircrafts and study the gases flow in high-enthalpy thermal chemical reaction, such as conventional hypersonic wind tunnels that obtain high Mach numbers by reducing the static temperature, and hypersonic wind tunnels that directly heat air and high-enthalpy shock wave tunnels that apply strong shock wave compression. These wind tunnels have made great progress with different methods to improve the Mach number, total temperature and flow velocity of the test airflow, thus providing a lot of data for hypersonic flow and aircraft design. However, the test airflow they provide is also restricted to varying degrees due to the limitations of various wind tunnel driving capabilities and heating methods, which makes it difficult to fully reproduce the flow field conditions under actual hypersonic flight conditions, such as flight Reynolds number, flight Mach Number, total temperature, total pressure, total temperature/wall temperature ratio, density ratio before and after shock wave, flow field pattern, geometric dimensioning of test model, etc. Even the most advanced hypersonic wind tunnel around the world can only partially simulate part of flight conditions, so the results obtained from ground wind tunnel tests cannot generally be directly applied to aircraft design and flight test planning.

What is more difficult is that there are differences in the flow conditions that different types of equipment for hypersonic ground test can simulate, which in turn leads to differences in test data. For example, the test conditions between different wind tunnels under the same Mach number also have significant differences in key parameters such as Reynolds number, static temperature, total pressure, and flow speed, which makes the obtained wind tunnel test data short of comparability, which brings great uncertainty on the prediction of aerodynamic performance and the study of the gas flow law of high-enthalpy thermal chemical reaction. Therefore, it is of great significance to understand the laws of high-temperature gas flow to improve the reliability and applicability of wind tunnel test data with the research on the theory and corresponding methods of hypersonic wind tunnel test data, and further obtain the prediction method of the consistency of the world from the ground wind tunnel test data to the flight conditions for the development of hypersonic flight technology.

It is required to carry out a lot of simulation and test work for the aerodynamic thermal environment research, which involves a large number of test equipment, test model resources (test template, test environment model, test variable complexity model, commercial software model, self-developed software model, etc.), simulation and test data, while the storage of such equipment and data is scattered, which is difficult to view and find in a short time. Therefore, it is necessary to effectively manage these resources, simulation and physical test data through establishment of a unified, standardized, safe and convenient spacecraft data mining system, which can provide on-line storage and query of resources and data by component and version, data real-time viewing and client-side lightweight display, data pre-processing and result analysis, and provide a variety of data mining algorithms to achieve data exchange and sharing between heterogeneous and distributed resources, and to achieve effective storage and management of resource data, models, and knowledge, thus making full use of these resources to achieve the goal of combining virtual and real, and applying the conclusions of data analysis to the overall plan design level, in order to provide strong support on the development.

2.2 A Review: Ground-Flying Consistency Research of Hypersonic Vehicle

Aerodynamic force performance and aerodynamic heat protection are two key factors that are required taken into consideration in the design of new aircraft.

However, it is difficult to fully simulate the inflow parameters and model status under flight conditions for ground tests under hypersonic conditions. For example, the inflow parameters and the model shrinkage ratio result in certain differences between the aerodynamic/thermal test results and flight results, which leads to a compromise in design. Therefore, the research on the prediction and verification methods of aerodynamic force/heat consistency between ground and flying is the key technical breakthrough point for the successful development and performance improvement of new aircraft.

People have successfully designed subsonic and supersonic aircrafts in the development process of aerospace aircrafts for more than 100 years, and applied a lot of research work on the correlation of wind tunnel test data with similarity theory, in which it obtained many engineering calculation methods with practical application values. However, it has not really obtained a similar theory suitable for hypersonic wind tunnel test data over 60 years of research on hypersonic flight technology, which makes the processing of hypersonic aircraft aerodynamic force/heat wind tunnel test data be still limited to empirical interpolation and extrapolation correction. In fact, there is no reliable similarity criterion for hypersonic ground tests due to the effect of high-temperature gas effects, and the extreme requirements of hypersonic flight conditions make the wind tunnel test simulation technology unable to meet the needs of actual flight conditions, which results in predicted aerodynamic characteristics Lack of reliability. The insufficiency of the research on the consistency between the ground and flying of the aerodynamic force/heat data of hypersonic aircraft has severely restricted the comprehensive application of a large number of wind tunnel test data, reduced the reference and guidance for aircraft design, and increased the cost, cycle and risk of development of new aircrafts. The continuous failure of the hypersonic aircraft X-51 and THV-2 flight tests in the United States highlights the complexity of hypersonic flow phenomena and the desirable demands for reliable prediction of aircraft aerodynamic characteristics.

There have been many published articles as the researches on the aerodynamic force/heat correlation started as early as the 1960s in the United States and Soviet Union and have been used for the heat transfer of rocket nose cones, nozzles and various internal combustion engines. It is focused on two aspects for the researches: On the one hand, it is the method of parameter identification solution, which takes the heat conduction equation as the state equation system for the parameter estimation, derives the state equation estimation parameter to obtain the sensitivity equation system and solves the state equation system and the sensitivity equation system with the difference method. On the other hand, it is the sensitivity analysis of measurement errors. AEDC from the United States tested the Rockwell60-OTS space shuttle model of the reduction model 7: 400 in the A and DC wind tunnels of AEDC under the test conditions of Mach 4, total temperature of 411 K, compared the test results, carried out analysis on the aerodynamic thermal characteristics of the surface interference zone, non-interference zone and main protrusions, and extrapolated the wind tunnel aerodynamic heat measurement results to the actual flight through correction factors. The United States spent about 90,000 h of wind tunnels researches on the development of the space shuttle, with use of 44 wind tunnels, including 5 m transonic, supersonic wind tunnel, 1 m hypersonic wind tunnel, 2 m high-temperature heat structure wind tunnels, 1.5 m piston wind tunnels in high Mach number and variable Reynolds number, etc., however, they still cannot meet the requirements of similar parameters, and have to correct the parameters that have not been simulated. The main amendments included at that time, structural deformation, flow field parameters, the shape resistance caused by the roughness of heat protection materials and protrusions, and changes in flow field parameters. The data was organized to form the ADDB database gradually. The United States carried out a series of flight tests at the same time and the tests were divided into two types: model and full-size, during which a large amount of data was accumulated and a basis was provided for modifying existing theories and ground test results to form OADB.

The research on wind tunnel test data association theory, association database and the algorithm of conversion between ground and flying is very critical to the design of aircraft and the prediction of flight data. For example, the problem of the trim angle of attack of US Space Shuttle is caused by the mismatch between the trim angle of attack predicted by the database and the flight test. Another example, the X-43A in the United States failed due to rudder damage in the entire flight test during the first flight.

It was found in analysis that the cause was that the inaccurate ground test data resulted in inaccurate prediction of flight aerodynamic force, which affected the design of the rudder shaft. They had to conduct ground test research again after the first flight of X-43A failed, revise the database, and redesign the rudder according to the revised results, thus ensuring the success of the second and third flight tests. It indicated in the experience and lessons learned by the United States in the development of hypersonic vehicles that it is important for the research on the association theory of hypersonic wind tunnel test data, test database and the ground-space conversion algorithm.

The Soviet Union and Russia have made significant contributions in the field of aerospace, including a great amounts of wind tunnel tests during the research and development of the space shuttle buran, and supplemented by theoretical analysis and flight tests at the same time to form a database can the initial program be determined by analysis of the aerodynamic shape database, thus saving a lot of money. Royal Aircraft Establishment in the United Kingdom has built a wind tunnel test and analysis system for individual components and assemblies of tactical bombs, which is taken to coordinated with the aerodynamic calculation of high angle of attack to complete the aerodynamic analysis of tactical bombs.

Most of the trajectories of spacecrafts are in the thin atmosphere or the outer atmosphere, in which it is inevitable to encounter the coupling problem of the thinning effect, the viscosity effect, the compression effect and the high temperature real gas effect. It has developed advanced numerical calculation method, wind tunnel test method in the ground and flying conversion methods of the aerodynamic characteristics from abroad, and the ground design method was improved through the modification of a large number of flight data. There is a high level on the theoretical basis of the ground and flying consistency of aerodynamic characteristics, ground workload and ground and flying conversion experience.

From the perspective of the current research status of the simplified contour, especially the Moss hollow column-skirt, it has achieved preliminary results in foreign researches, and it has started to serve the research on aerodynamic characteristics of the hypersonic rare transition zone. The aforementioned various gas effect corrections could only give the effect of the overall characteristics in the 1970s, however, they can already give the effect of the local characteristics at present, that is, the effect of various effects on the aerodynamic heat distribution.

It started early in the management of experimental data in other countries, and there is a good technical foundation for the research of experimental data mining technology. Big data technology has also been continuously developed and applied in many fields abroad with the upsurge of big data around the world.

A mature evaluation mechanism has been formed abroad in the aspects of the environmental effects of space radiation, and rich research results have been worked out for the spatial environment model and the environmental effects of space radiation. NASA, ESA, French Space Agency and Russian Space Research Agency have developed their own space radiation environmental effect analysis software, which includes both commercial software and on-line analysis software.

The research progress of the lean gas effect in China is relatively slower compared with other countries, which focused on numerical simulation mostly, and does not have mature ground test conditions that can reflect the lean gas effect in the flight environment. The research on non-equilibrium fluids in China mainly focuses on theoretical analysis and numerical simulation. There are few domestic studies on the physical and chemical models of non-equilibrium effects and wall catalytic effects due to insufficient wind tunnel test capability and lack of first-hand test data, lack of test basis, and comparative verification with test results is also less, which mainly relies on ground test and flight test data from abroad. In addition, the corresponding model is more difficult to implement in the numerical method due to the complexity of the theoretical model of thermochemical non-equilibrium.

The more mature general computing software in China similar to foreign LAURA needs to be further developed.

There are still some limitations in the field of data management and integration of self-developed systems in China. The development and application in the field of test data management is very rapid. There have been commercial test data management systems appeared on the market at present, such as Newtera TDM and TDM3000, etc., however, it is relatively lacking for the analysis algorithm for mining massive test data.

China has carried out a lot of ground test research on flying aircrafts. The aerodynamic heat problems have been well resolved on powerful launch vehicles, return satellites, and manned spacecraft return modules, which has laid a good foundation for the research of the aerodynamic force and heat consistency between the ground and flying. Hypersonic technology has achieved some results after nearly 20 years of development, but there is still a certain gap in the comparison of hypersonic technologies with the leading countries of the United States, Russia, etc. It already has a powerful means of conducting aerodynamic heat ground test research on hypersonic aircraft in terms of aerodynamic force/heat theories, calculations and experimental research of hypersonic aircraft, and has established a systematic test method and management system. The main problem is that the test equipment at present cannot be demonstrated to verify the aerodynamic force and aerodynamic heat tests of the full-scale model of the aircraft under real flight conditions. It is difficult to carry out research on the aerodynamic heat distribution and heat load of the aircraft. At the same time, the technique of hypersonic flow measurement also needs to be improved There is no systematic and comprehensive method for predicting the aerodynamic characteristics of the world in terms of aerodynamic thermal environment prediction, and related methods and tools are in urgent need for verification and improvement of flight data.

2.3 Development of Method for Conversion of Aerodynamic Heat on Ground-Flying

2.3.1 The Aerodynamic Heat Difference Between Ground and Flying

It is the key technology for flight success or failure to the design the heat-preventing system of high-speed and complex-shape aircraft. It is the basis of heat protection design to determine the aerodynamic heating of the incoming aircraft from outside. The accuracy and precision of energy processing, especially for long-time, large heat-load flight, directly correlate and threaten the reliability of heat protection design. There has been pneumatic heating engineering and numerical calculation methods, pneumatic hot wind tunnel ground test measurement technology, arc heater material ablation and heat structure simulation test. The hypersonic wind tunnel aerodynamic test is still the main method at present to study the prediction of the aerodynamic performance of hypersonic aircrafts and to solve key aerodynamic problems (1). It is necessary to establish the correlation between aerodynamic heat ground test and flight conditions, that is, the consistency of aerodynamic heat between ground and flying, so as to apply theoretical analysis and numerical simulation, ground test to flight conditions. The aerodynamic heat consistency between ground and flying is to study the difference between the real flight environment and the ground simulated environment, establish the correlation method of aerodynamic heat environment data between ground and flying, deduce the wind tunnel test results to the flight conditions, and predict the aircraft flight thermal environment. It mainly solves the law, trend and reliability within the range of heat environment between ground and flying.

The heat environment of the aircraft is subject to incoming flow conditions (including incoming Mach number, Reynolds number, angle of attack and angle of sideslip), surface flow field characteristics (including laminar flow, transition (2) and turbulence), surface material characteristics (including surface roughness, surface oxidation resistance, surface adsorption characteristics, etc.). It is difficult to fully simulate the above many influencing factors whether it is ground test or numerical simulation, which brings great difficulty to the consistency of the ground and flying test, therefore, it is very important to carry out correlation analysis of thermal environment data.

The hypersonic aircraft with complex contour adopts a complex contour, large flight speed, wide range of altitude, and large changes in environmental parameters. Therefore, the factors affecting the aerodynamic thermal environment are very wide and complex, including almost all of the pneumatic problems for large launch vehicles, manned spacecraft, and reentry aircraft recovery. The aircraft often passes through the dense atmosphere and the thin gas layer when it enters high altitude flight, that is, the continuous flow area, the transition area, and the free molecular flow area (usually it is divided through the Knudsen number [Knudsen] the flow area, and different flow areas use different calculation methods), and there are different flow states: laminar flow, transition and turbulent heating, so it is more difficult to calculate the aerodynamic heat environment with a single method. Therefore, different calculation methods are used for different shapes in the design of the aircraft. The aerodynamic heating of the aircraft is very complicated and has many influencing factors even if the same shape and different flying heights require different calculation methods at different positions of the aircraft. The reliability and feasibility study of pneumatic heating must be carried out, so as to meet the requirements of heat protection design.

The aerodynamic heating data required for the design of hypersonic complex contour are obtained from reliable calculation methods and from ground test results. It often requires ground test verification even if the calculation method is very mature. It depends to a large extent on the ground test results if the calculation method is not mature. However, it is often impossible to meet all the test simulation criteria of actual requirements due to the limitation of test equipment (or method), so some simulation criteria are relaxed or abandoned. It requires solving the following two problems: ① What extent can the obtained test results be applied to the real flight state; ② What is the theoretical analysis and calculations that should be carried out on the test results in order to extrapolate the existing test results to actual flight conditions. Both of the two parts of work are called correlation research on aerodynamic heat between ground test and flight test, and also called aerodynamic heat "Ground and Flying Consistency". The re-entry heat environment and the total heat load are more severe as the range and re-entry speed increase according to the needs of development, especially the local heat protection parts. The development is facing very difficult problems for the heat protection system. The influence of the heat environment on the material ablation must be considered in order to complete the extrapolation design of the aircraft structure and thermal protection system. Therefore, it mainly considers the changes in the ablation profile caused by the aerodynamic thermal environment to determine the basic parameters and simulation criteria of the ground simulation test for extrapolation ablation. It is believed in the analysis that the difference between the ground simulation test and the flight status in the air lies in:

(1) Ground tests cannot fully simulate the heat environment in the flying;
(2) Scale effect, the influence of scale model;
(3) The impact on the quality of the heat environment flow field.

But can the ground test results be directly used in flight status? What modifications are required if the ground test can be used for flight status? It requires quantitative processing of ground test results to find out the characteristic geometric parameters that meet the regularity as the basis for the ablation profile in the flying. Usually the key parameters are analyzed to find the difference between the ground test and the flight state. It is believed after analysis that there are many calculation equations that affect the contour of ablation, which can be summarized as follows:

(1) Calculation equations for different pressure;
(2) Calculation equations for different heat flow;
(3) Calculation equations for different heat increment;
(4) Calculation equations for different transition;
(5) Calculation equations for different mechanical ablation;
(6) Calculation equations for different thermochemical ablation;
(7) Calculation equations for different ablation profiles;
(8) Calculation equations for different shock wave contours.

It is believed after years of research that the above eight items that affect the ablation profile have been basically unified, and the key parameters public recognized include roughness, laminar flow, transition, turbulence, and transition Reynolds number. Mechanical ablation, thermochemical ablation, and ablation profile calculations among the above eight items have little to do with the calculation of cold wall heat flow. The remaining items are issues that must be studied in calculating cold wall heat flow. The successful simulation flight test assessment results are extrapolated to the extended range flight trajectory by theoretically calculating the heat environment and ground simulation test after the extended range, so as to complete the extrapolation design of the overall structure of the warhead and the heat protection system.

Over the years, the progress of research on consistency of the aerodynamic heat between ground and flying has been slow due to the lack of flight data. More emphasis has been placed on overall parameters although some data have been measured in the past for large launch vehicles, return satellites, and manned spacecraft return cabins. There is no detailed heat flow data for heat environment data, which brings difficulties to the consistent study of ground and flying. The study can be carried out on the law of the consistency of the ground and the flying with the ground test, flight test and simulated data after the obtainment of detailed aerodynamic thermal data through special aerodynamic heat flight test data and combination with numerical simulation research at the same time, which focuses on the influence of the deviation input. Finally, the distribution law of the heat environment is proposed, which is the development direction of the research on the aerodynamic heat consistency between ground and flying.

2.3.2 Data Correlation of Aerodynamic Heat Tests

The differences between the wind tunnel test and the real flight test include flow patterns, incoming flow quality, tunnel wall and bracket interference, wall temperature and humidity, measuring instruments, aerodynamic elastic deformation, model fidelity and installation, real gas effects, etc. The wind tunnel test results need to be revised in order to reflect the real flight situation by the wind tunnel test. Praharaj et al. [1] researched on the scale effect between the supersonic missile jet ground test and real flight, and verified the reliability of the correction method of the ground test scale with CFD numerical calculation. Ewald [2] conducted a summary research on the correction method of the wind tunnel wall and collected and organized the commonly used wall correction methods. Viswanat and Rajendra [3] and Cahn [4] carried out analysis on the influence of model poles on ground wind tunnel test results and developed corresponding correction methods. Steinle and Stanewsky [5] conducted a research on the impact of inflow disturbances in the wind tunnel on the test results.

The researchers in China have also carried out corresponding research, Ni and He [6] adopted equivalent dynamic pressure method and wall pressure information method to modify the wall interference of wind tunnel test data of Y-8 aircraft with propeller model. Yin et al. [7] took the measured cave wall pressure as the boundary condition, calculated the flow field with the Euler equation, compared it with the calculation result of the boundary condition being the far field pressure, and then it shows the method is feasible after the comparison with the test results. Zhang et al. [8] conducted numerical and experimental researches on the interference effect of the wind tunnel tail brace, and discussed and analyzed the impact of different support methods on the results.

Chen et al. [9] carried out prediction on the nonlinear aerodynamic characteristics of the aircraft with small aspect ratio and has completed the correction of the effects of brackets, cave walls, air intake, jet flow, Reynolds number and static elasticity. Yan et al. [10] studied the influence from different calculation formats on the accuracy of heat flow calculations, and believed that the AUSM + format is the most accurate for heat flow calculations. Pan et al. [11] discussed the influence of reseau form on heat flow calculation. It was shown in the results that the heat flow calculation is extremely sensitive to the wall normal grid distribution due to the large wall temperature gradient and it was found that the convergence of aerodynamic heat calculation is much slower than the pressure and flow field convergence, that is, it is incorrect to take the equation residuals and pressure convergence as the basis for aerodynamic heat convergence. Zhang et al. [12] carried out analysis on the influence of different turbulence models on the calculation of heat flux density, and it was found that the SST model and the k-ω model have good results in the calculation of heat flux density after the comparison with the experimental results.

Such work for in-depth analysis such as extrapolation of aerodynamic thermal test data is continuing and deepening in various countries at present. For example, the hypersonic demonstration and verification flight plan from NASA in the United States has been carried out for many years, and it has fully demonstrated the research and development of hypersonic vehicles from the activities of emergence and off-shelf of various types developed, such as X-30, X-33, X-34, X-37, X-38, X-43, X-51, etc.

NASA of the United States has vigorously developed hypersonic aircrafts to carry out global strikes that is in long range, precise and rapid, and has conducted in-depth analysis and full evaluation. This is a good reference for the research and design of hypersonic aircrafts in China.

As it believed in the research on the mechanism of aerodynamic heat consistency between ground and flying that there are two types of ground simulation tests in the aerodynamic and heat design of aircraft: one type is simulation tests on aerodynamic performance based on similar criteria.

The physical quantities simulated are dimensionless aerodynamic coefficients, such as lift coefficient, drag coefficient, torque coefficient, etc., no one doubts that the test results are directly pushed to the flying since the similarity criterion is strictly derived from the basic equations of aerodynamics. Therefore, the problem of the consistency between heaven and earth is not put forward in the aerodynamic design of the aircraft. The other type is taking the physical characteristics of the heat protection simulation test as basis, and the physical quantities simulated are dimensional quantities, such as ablation speed, surface temperature, etc. These tests are the analysis and research on the mechanism of physical quantity change in the ablation process based on simulation theory. As the ablation of real materials can only be carried out under the condition of high heat flow, the absolute value of heat flow is the most critical environmental parameter to simulate the whole process of ablation. The heating time is also an important factor since the heat response has a hysteresis effect.

There are three factors that determine the size of the heat environment, namely the total enthalpy, pressure and simulated geometry. The total enthalpy is the most important factor among these three factors. The arc heater used in the thermal protection test is limited by the power at present, so the above three factors cannot be met at the same time, and only the most important one or two conditions can be selected according to the ablation and heat response mechanism. For example, it can select heat flow and total enthalpy as the main heat environment simulation conditions in stagnation thermochemical ablation test, while geometric dimensions and pressure are not considered as simulation conditions; and heat flow and pressure can be selected as the main thermal environment simulation conditions if considering mechanical ablation of materials; the geometric size is an important factor for the ablation shape test, so the largest possible size and heat flow conditions are selected. A question naturally arises since the arc heater cannot fully simulate the conditions of air heat flow that, can the test result of the arc heater be pushed directly into the air, that is, the problem of consistency between the ground and the flying. This problem was mainly solved through theoretical analysis and calculation methods over the past. For example, the macro heat protection theory basically solved the heat protection parameters in traditional design, but the ablation profile involved in the new aircraft is much more complicated than determining the heat protection parameters.

There is still a certain gap from the foreign technology on ground and flying conversion of aerodynamic heat and the refined design of the reusable hypersonic aircraft heat protection system although more systematic research work has been carried out on the technology on ground and flying conversion of aerodynamic heat, which is mainly reflected in the following aspect:

(1) There is still some manual intervention in the selection of heat flow conversion equations for large areas of different types of aircraft in the case of laminar flow;
(2) Technology of large-area area aerodynamic heat conversion of ground and flying under turbulent conditions still needs more verification work;
(3) There are few researches on the influence of wall temperature ratio on aerodynamic heat conversion technology;
(4) The heat environmental conversion technology between ground and flying in the local complex flow area has only carried out a preliminary research on the leading edge of the wing rudder, and has not conducted research on the compression of corners, slits and other complex flow structures;
(5) All the above researches are based on the ideal gas, and it needs to be further studied for the effectiveness and reliability of the method of aerodynamic heat conversion between ground and flying under the effect of high-temperature real gas.

2.3.3 Method of Aerodynamic Heat Conversion of Environment in Large Area

The research on the conversion method of aerodynamic heat environment in a large area mainly includes the following four aspects.

1) Research on Selection of Sensitivity Parameters

Different types of aircraft or different parts of the same aircraft surface require different conversion methods for correlation extrapolation of the aerodynamic heat test data, and the selection of correlation equation is closely related to the compressibility of the local airflow.

Selecting the correlation equation based only on the Mach number at the outer edge of the local boundary layer often fails to obtain a good correlation result although the Mach number can be used as a sign of airflow compressibility. Therefore, the selection requires experience of the conversion equation for the specific aircraft contour. The most sensitive parameters for shape dependence can be selected for different types of aircraft, such as the contour of a large blunt head, the contour of a flat body, and the contour of a slender body, etc., and then take it as a method for selecting and converting a large area of aerodynamic heat area and basis for conversion equation correction.

Analysis has been carried out on the variation of the dimensionless heat flow coefficient q/q_s, Stanton number St, and heat transfer coefficient h with the geometric scale of the model, the Mach number, the Reynolds number, the total temperature and pressure of the flow field according to different flow states and flow characteristics, such as adhesion flow, separation flow, shock wave interference flow and other flow characteristics in different regions through theoretical analysis of the heat flow equation of the similar solutions of the flow control equation and the boundary layer equation, so as to obtain the main aerodynamics that affect the aerodynamic heat conversion parameters and geometric parameters. CFD and engineering calculations are carried out for typical wind tunnel tests and flight conditions based on the theoretical study of sensitivity parameters. The influence of various parameters on aerodynamic heat is analyzed and studied based on the theoretical calculation results.

A ground wind tunnel test verification study was carried out accordingly: different scale model heating tests were carried out on the shock wave wind tunnel according to the different types of aircraft contours. Analysis and research was carried out on the influence law of various sensitivity parameters, such as scale, Reynolds number, total temperature, total pressure and multi-parameter coupling effect, on aerodynamic heat according to the different flow characteristics of the attached flow, the separated flow and the shock wave interference flow based on the wind tunnel test results, and the theoretical analysis results were tested and verified.

2) Research on aerodynamic heat conversion method under turbulent conditions

The main method of theoretical research on the correlation of aerodynamic heat environment between ground test and flying test under turbulent conditions is to take engineering calculation equations as tools to derive and analyze the differences in heat flow distribution under different inflow conditions, scientifically simplify the secondary factors in the heat flow equation, eliminate similar factors by similarity, and finally simplified to obtain the correlation factors of the wind tunnel test in ground conditions and flight conditions. And then establish a preliminary extrapolation prediction theory, the derivation processes the rationality of approximation and simplification is analyzed through a large number of wind tunnel test results and flight results verification, and finally the correlation theory of reasonable heat environment ground test and flight conditions is obtained.

Researchers have developed different calculation equations for aerodynamic thermal engineering calculation equations under turbulent conditions. The heat flow calculation equations are different, and the calculation results are also somewhat different, sometimes even different by 20–30% due to the difference between the approximate assumption and the reference enthalpy used in the solution process.

Therefore, it is required to analyze and compare the differences of different turbulent heat flow calculation equations, so as to obtain a more reasonable and reliable method for aerodynamic heat conversion between ground and flying.

The focus at present is on the difference between different turbulent heat flow calculation equations and the application range of different heat flow equations. On this basis, the derivation of each heat flow equation is simplified, to establish aerodynamic heat conversion method under turbulent conditions, and compare for the difference between different conversion methods.

Correlation parameters are established for the analysis of different turbulent heat flow equations, and research is carried out on differences and applicability, especially for the wide range of Reynolds number change, the range of incoming static temperature change, and the package under wind tunnel test and flight conditions, which includes the influence on the scale of the model.

The factors of various influencing parameters established by the correlation theory analysis are corrected to truly establish the different parts of the complex shape and different flow characteristics. Related conversion parameters through a large number of experiments and numerical simulation results.

3) Wall Warming Effect Research

The aerodynamic heat correlation parameters and the ground and flying conversion method given in the current study are based on the assumption of cold wall heat flow. The wall temperature changes with time during actual flight, which results in the wall temperature of the model surface under wind tunnel test being quite different from the wall temperature on the surface of the aircraft under flight conditions, and it is necessary to carry out analysis and research on the influence of wall temperature or wall temperature ratio.

A preliminary research was carried out on the effect of the wall temperature of the model surface on the conversion results in the case of laminar flow. It is shown in the result that the conversion results are greatly affected by the wall temperature ratio of the aircraft stagnation point and the local wall temperature ratio of the aircraft surface. However, there is no research at present on the effect of wall temperature ratio under turbulent flow, so it is necessary to conduct in-depth research on hot wall correction and wall temperature ratio effect for the current conversion method.

The effect of wall temperature ratio on the heat environment of the surface is more significant as the total temperature of reusable aircraft is relatively low when reentering. The wall temperature of the current wind tunnel test is about 290 K, the total temperature is in 430 ~ 2280 K, and the wall temperature ratio varies from 0.12 to 0.67, which is very wide, making it difficult to carry out variable wall temperature tests. Therefore, the research is estimated to carry out from two aspects: one is to study the difference in the heat environment of the aircraft surface under different wall temperature ratio conditions with the engineering calculation results and numerical calculation results, and to verify change of the large area regional cold wall heat flow, complex flow area cold wall heat flow (stagnation point/leading edge, compression corner, rudder wing area, etc.) with wall temperature ratio based on the existing wall temperature ratio impact research, and establish the relationship between the wall temperature ratio and the surface thermal environment of the aircraft with comprehensive investigation on its physical mechanism; the other is to select the typical test state and shrinkage model for the wind tunnel heat test for the typical flight tests at home and abroad, according to the shock wave wind tunnel test conditions, the surface heat flow in typical flight state is predicted through extrapolation of the wind tunnel test data conversion, including cold wall heat flow and hot wall heat flow with consideration of the effect of wall temperature, and carry out engineering and numerical calculation on the flight state. The conversion method considering the wall temperature correction is improved and verified by comparing and analyzing the prediction results, calculation results and flight measurement results between ground and flying tests.

4) Research on high-temperature real gas effect

The real gas effect of high-temperature gas on the aerodynamic heat of the aircraft has caught attention from the aerodynamics community with the development and promotion of reusable hypersonic aircrafts. One of the important issues is the experimental simulation of real gas effects, for which it is necessary to understand the similarity of the flow of real gases. However, the current similarity law of high-temperature gas, the dual scaling law, is obtained under the collision reaction conditions of two bodies, which is not the case with real high-temperature gas. Therefore, it is essential to carry out research on the flow of high-temperature gas with chemical reaction and corresponding similar criteria.

The flow accompanied by high temperature causes the air molecules vibrate, dissociate or even ionize under the conditions of hypervelocity flow, which makes ordinary air a complex medium that is constantly undergoing thermochemical reactions. Real gas effect is the outstanding characteristics of the hypersonic flight environment. Research on aerodynamic heat focuses on the thermochemical reaction flow, including the thermochemical changes of the gas flow itself and the catalytic and cracking reactions between the gas flow and the model surface.

It is required to meet gas composition, total airflow temperature, airflow velocity, and characteristic density to fully obtain aerodynamic heat researches with real gas effects, ground test simulation equipment. In addition, it must guarantee the Damkoler number that is related to the flow characteristic time and the chemical reaction characteristic time. Researches shall be carried out from the following three aspects: the first is to conduct research on the scope of application of the current dual scaling law; the second is to study the dependence of each physical quantity on the non-equilibrium effect in the hypersonic non-equilibrium flow, and the third is to derive similar criteria for non-equilibrium flow from the high-temperature air non-equilibrium flow NS The equations and boundary conditions, which are important for both forward and reverse reactions, and take focus on how to convert the results of aerodynamic heat tests on ground wind tunnels (including complete gas and high-temperature real gas) into high temperature equilibrium heat environment prediction method under equilibrium flow.

2.3.4 *Conversion Method of Local Aerodynamic Heat Environment*

The current research methods can not accurately obtain the value of pneumatic heating due to the complexity of local problems. There is a certain similarity of the conversion method between ground and flying with that for large area, but it also has its uniqueness, and it must be specially studied.

2.3.4.1 Correlation Between Aerodynamic Heat of Ground and Flying in Local Complex Interference Area

There are serious shock waves/shock wave interference, shock wave/boundary layer interference and other complex local interference phenomenon of aerodynamic heating in the case of high Mach number, which makes the heat flow of the wing leading edge, local interference area, rudder surface and other parts very high. The local heat environment often determines the success or failure of flight.

The change curve of dimensionless heat flow with Re was obtained through the fitting of wind tunnel test data for typical local interference areas, such as leeward separation area on the surface of the shuttle, eddy current up-current area on the side, fairing and other local interference areas in the process of developing the space shuttle in the United States, which is used to directly extrapolate to flight conditions on the premise of the same angle of attack. The flight data is consistent with the extrapolated data from the wind tunnel test within the Re range of the wind tunnel test.

Comparison between temperature values and flight test data is made from the prediction on control points on the elevons. The predicted value of the original model tends to be conservative, and the results after adjustment through the flight test data are in good agreement with the actual flight results, but the transition under actual flight conditions occurs at a higher Reynolds number (that is late flight) compared with the predicted results.

Comparison between temperature values and flight test data is made from the prediction on control points on the body of the flap.

The initial heating model only included the situation that the body flap declination is less than 15°, which is to extend the model to the situation that the body flap angle is 22.5°, which added wind tunnel data to the heating model after flight adjustment. The flight data indicates that the flow on the body flaps may be in a transitional phase during most of the reentry phase.

The angle of attack, yaw angle between vortex position and intensity and Re sensitive data is derived from wind tunnel test data for the vortex up-current area on the side of the space shuttle. The changes of these parameters in the wind tunnel result in that the vortex moves forward or backward along the side of the fuselage. The dimensionless heat flow is not sensitive to the change of Re for the side of the front fuselage, and moves backward along the side of the fuselage, at 30% of the length of the fuselage. The flight data and the extrapolating data of wind tunnel test is agreed within the Re range of the wind tunnel test.

The flight data is lower than the extrapolated value for the case of lower Re, while for high Re, the flight data is generally equal to or slightly higher than the extrapolated value. The difference between the wind tunnel extrapolation data and the flight test data is large for regions strongly affected by vortex washing, and this difference increases with the increase of Re.

The extrapolation of the wind tunnel test data of vertical tail can be divided into vortex-free areas and vortex-related areas. Generally speaking, the flight data and the wind tunnel test extrapolation data are basically the same for the vortex-free area, which is mainly used in the lower half of the vertical tail. The flow is similar to that on the side of the fuselage for the area affected by the vortex. The flow is separated at low *Re*, and the flight data is lower than the wind tunnel extrapolation results. The data under flight conditions rises rapidly when $Re = 2.0 \times 10^6$, which makes the flight data consistent with the wind tunnel extrapolation data or slightly higher at higher Re [13].

The flow characteristic is that a pair of weak vortices pass along the top of the door in the wind tunnel test for the position of the payload door, but the pair of weak vortices are at a 40 in the flight test. It does not appear to be adhesion at high angles of attack. The relevant parameter used is the Stanton number based on free flow conditions in order to extrapolate the wind tunnel data of the payload door. As the angle of attack of the aircraft decreases, the gap between flight data and wind tunnel test extrapolation data also decreases when this parameter is used for wind tunnel data and flight data.

Foreign countries have also conducted a large number of wind tunnel test studies for the aerodynamic heat problem in local areas on typical flow models such as compression corners, forward steps, protrusions, and cavities by simplifying the contour. The test data are passed and the relationship between peak heat flow and peak pressure is established on this basis:

$$\frac{\dot{q}_{peak}}{\dot{q}_{in}} = \left(\frac{p_{peak}}{p_{in}}\right)^n \tag{2.1}$$

In the equation:

q_{peak}—peak heat flow (kW/m^2);

q_{in}—heat flow (kw/m^2);

p_{peak}—peak pressure (Pa);

p_{in}—local pressure (Pa).

The calculation equation for local heat flow in the typical flow area is used for wind tunnel test and calculation and prediction on aerodynamic heat of flight conditions. This method established to predict the flight thermal environment through the extrapolation of the wind tunnel data based on theoretical analysis and experimental data has been successfully applied in the design of the flight heat environment of space shuttle orbiter and has been verified by the flight data. In particular, the prediction results were analyzed and compared for the different feature points on the surface of the space shuttle after obtaining a large amount of flight measurement data, and the method was further revised to aerodynamic heat prediction of subsequent aircraft under flight conditions of X-34, X-38, X-37B, HTV, etc., which provides a solid technical foundation.

Comparison was carried out between the extrapolated predicted temperature of a lower surface line and the CFD results when X-38 aircraft at a Mach number of 6, 40° angle of attack, 20° rudder deflection angle. The extrapolation results of the wind tunnel test for large-area interference-free areas are in good agreement with the CFD calculation results. The extrapolation results are higher than the CFD calculation results for the rudder surface interference area. Generally speaking, both of the results agree well [14].

However, it is worth pointing out that the main basis for the establishment of this method is the ground wind tunnel test data of the space shuttle contour, and the method is corrected by the space shuttle flight data, therefore it is needed to carry out further verification for universality of other contours.

Preliminary research has been carried out on the aerodynamic heat conversion method of the local thermal environment over the recent years, which is mainly for the rudder leading edge shock wave/boundary layer interference, shock wave/shock wave interference, analysis of the characteristics of the flow field structure. Starting from the calculation equations of the heat flow of different rudder leading edges after taking into account the factors of the Reynolds number, Mach number, the main factors are preliminarily obtained that influence the ground test and flight status of the aerodynamic heat environment of the wing leading edge and rudder surface.

The heat transfer relationship under laminar flow conditions between the leading edge of a swept wing and the leading edge of a non-swept cylinder is:

$$q = q_{\Lambda=0} \cos^n \Lambda \tag{2.2}$$

In the equation:

q—heat flow of the leading edge of the cylinder with sweep angle Λ (kW/m^2);

$q_{\Lambda=0}$—heat flow of the leading edge of the cylinder without sweep angle (kW/m^2);

Λ—Sweep Angle (°);

n—Constant, when Λ is in the range of 0–60°, It can be 1.5, and 1.0–1.5 in the larger sweep angle range.

In fact, there is the following relationship due to the heat flow of the non-swept cylinder and the heat flow of the ball head of equal radius:

$$q_{\Lambda=0} = q_0/\sqrt{2} \tag{2.3}$$

In the equation:

q_0—the stagnation heat flow under the same flow conditions as the ball head with equal radius without a swept cylinder.

Therefore, it is also written as follows in some references:

$$q = 0.707 q_0 \cos^n A \tag{2.4}$$

Cato presented another method for estimating the heat flow at the leading edge of the laminar wing:

$$q = 0.75q_0\left[1 - 1.857\left(\frac{\Lambda}{90^\circ}\right)^2 + 1.097\left(\frac{\Lambda}{90^\circ}\right)^3\right] \tag{2.5}$$

It can be seen that both of Eqs. (2.4) and (2.5) consider that the dimensionless heat flow of the leading edge normalized to the stagnation point heat flow of equal radius under the laminar flow condition has nothing to do with the incoming flow condition, which only depends on the effective sweep angle. The equivalent sweep angle is also the same for the same contour under different conditions, if the angle of attack is the same. The ratio of the radius of the leading edge of the wing to the radius of the stagnation point of the aircraft head is equal for different reduction ratio models or full-size aircraft, and the effective sweep angle is the same under the same angle of attack. Therefore, the dimensionless heat flow with reference to the stagnation heat flow is irrelevant to the model scale, Mach number, and Reynolds number for the leading edge of the wing, without considering the shock interference. The dimensionless heat flow of the leading edge of the wind tunnel test is equal to the flight situation in the case of the same angle of attack [15]. That is:

$$\frac{\left[\frac{q}{q_0}\right]_2}{\left[\frac{q}{q_0}\right]_1} = 1 \tag{2.6}$$

However, Johson gave the equation for calculating the heat flow of the leading edge under turbulent conditions:

$$q = 0.75q_0\frac{K_T}{K_L}\left[\frac{2\rho_\infty U_\infty R_N}{\mu_\infty}\right]^{0.3}\left[0.01714 + 0.01235\sin\left(3.53\left(\Lambda - 10^3\right)\right)\right] \tag{2.7}$$

In the equation:

K_T—turbulence factor;

K_L—Laminar flow factor;

ρ_∞—Incoming flow density (kg/m^3);

U_∞—incoming velocity (m/s);

μ_∞—Viscosity coefficient of inflow [kg/(m s)];

R_N—Cylinder radius (m).

The value of K_T/K_L is 1.414.

It can be seen from the equation that the dimensionless heat flow at the rudder leading edge normalized by the heat flow at the stagnation point of an equal radius ball head is not only related to the equivalent sweep angle, but also related to the Reynolds number of free flow, leading edge radius, turbulence ratio and other factors. Well, obviously there is a scale effect and a Reynolds number effect at the same time. The dimensionless heat flow at the rudder leading edge has a scale effect for different shrinkage models under the same incoming flow conditions. The local turbulence may be inconsistent even if the incoming flow parameters are exactly the same, so the turbulence parameter cannot be eliminated. In summary, the relevant parameters of the leading edge of the turbulent wing can be expressed as

$$\left(\frac{q}{q_0}\right)_2 \Big/ \left(\frac{q}{q_0}\right)_1 = (R_{N2}/R_{N1})^{0.3}(Re_{\infty 2}/Re_{\infty 2})^{0.3} f_b\left(\left(\frac{K_T}{K_L}\right)_2 \Big/ \left(\frac{K_T}{K_L}\right)_1\right) \tag{2.8}$$

It can be seen from the calculation equation of the heat flow at the leading edge of the wing under the conditions of laminar flow and turbulence above that, the influence parameters of heat flow under different flow conditions are quite different in the wind tunnel test and flight conditions, which needs to carry out calculation and related experimental verification on the derived correlation parameters, and further analysis and research are needed on the influence parameters based on other wing leading edge heat flow equations.

2.3.4.2 Similar Criteria for Aerodynamic Heat Environment in Local Areas

The research on the aerodynamic heat environment similarity criterion in the local flow area is mainly to analyze the physical mechanism controlling the local complex flow, to explain the characterization parameters of the physical mechanism and the characterization relationship of the physical phenomenon with combination of theoretical analysis and experimental verification, to establish the local complex The simplified flow model of flow, and the simplified model of the characterization equation, and to obtain the local flow aerodynamic thermal test simulation criteria according to the simplified characterization equation. There are many types of local problems on surfaces of different types of aircrafts, and there are also many differences between their structural characteristics and that of the aerodynamic heat environment. Researches are carried out on the critical local aerodynamic heat environment problems with serious aerodynamic heat environment on the surface of a typical shape aircraft, such as the compression corners formed by the deflecting rudder surface, the rudder shaft gap, the heat protection tile gap, protrusions and recesses the research.

1) Heat flow calculation model of typical local flow area

The research on simplified model of local aerodynamic heat flow is mainly to study the aerodynamic heat problems of deflection rudder surface compression corners, rudder shaft gaps, heat-resistant tile gaps, protrusions and cavity of typical shape aircraft, and to investigate similar research results from abroad of local flow models and the aerodynamic heat environment, including the results of the experiment and calculation of the local flow field structure and flow model, the theoretical model and calculation equation of the aerodynamic heat environment engineering calculation. Analysis was carried out on various local aerodynamic thermal engineering prediction methods and theoretical models, with combination of the local aerodynamic heat test of the ground wind tunnel to analyze the influence parameters of the local aerodynamic heat and extract the flow field parameters and their characterization relationships that control the local flow physical mechanism with theoretical analysis, so as to build a simplified model that can effectively describe different local complex flows, establish a theoretical model of local aerodynamic heat wind tunnel test simulation and engineering calculation, and evaluate the reliability and applicability of the simplified model with combination of the test and numerical simulation.

2) Research on influence parameters of local heat environment

The research on the influence parameters of the local flow aerodynamic heat environment mainly includes two aspects of contents: the first is to analyze and study the key parameters affecting local aerodynamic heat, including flow field parameters, flow state, boundary layer thickness, length of compression corner, height, relative height to boundary layer, width of gap, direction of gap, flow field around the gap and geometric characteristics with basis of the investigation of the local flow model, through comparing and analyzing various local aerodynamic heat environment engineering calculation methods and related equations, respectively for the compression corners and gaps, so as to analyze and study the influence of these parameters on the local heat flow and pressure distribution. The second is to verify the numerical and engineering calculations of typical wind tunnel test results at home and abroad based on the theoretical analysis results and the influence parameters of the local aerodynamic heat environment. In particular, research was carried out with the results of numerical simulation on the difference between the local aerodynamic heat environment and the results of full-model tests and local model tests with different reduction ratios under actual flight conditions. The influence law was analyzed and studied of local flow field parameters, boundary layer thickness, geometrical parameters of compression corners, gap width and surrounding flow characteristics changes caused by shrinkage ratio on aerodynamic heat of the gap. The key parameters and simulation criteria for wind tunnel test simulation of local aerodynamic thermal environment are determined for the compression corner and gap flow according to the research results of the above theoretical analysis and numerical simulation analysis, including simulation rules of Mach number, Reynolds number, boundary layer thickness, geometric scaling of compressed corners, geometric scaling of crevices, etc.

3) Research on the application of local heat environment ground test data

The research on application of extrapolation of wind tunnel test data for local aerodynamic heat environment should attach special stress on the consistency between ground and flying of the results of local aerodynamic heat test and the application of extrapolation of test data. The application of extrapolation of wind tunnel test data for local aerodynamic thermal environment is more difficult than the problem of large area, since even for the full model aerodynamic heat wind tunnel test with shrinkage ratio, there are two special difficulties for the aerodynamic heat test in the local area: the first is that it is difficult to make accurate heat flow measurements since the local geometric scale is very small after the shrinkage ratio. The second is that it may affect the local flow characteristics and peak heat flow due to the shrinkage ratio since the local separation and re-attachment of aerodynamic heat characteristics are often related to the local absolute geometric size and the relative thickness of the boundary layer. The local peak heat flux and the corresponding heat flux analog coefficient are very important for the application of local aerodynamic heat environment wind tunnel test data. Therefore, attention should be paid to the study of the influence of various flow field parameters on the peak heat flow and the corresponding coefficient of heat flow analogy in the application research of wind tunnel test data for local aerodynamic heat environment. The peak heat flow comparison coefficients especially obtained through experiments must have a clear correspondence with the comparison coefficients under actual flight conditions. For example, the focus is usually on the ratio of peak heat flow to local undisturbed heat flow for the local aerodynamic heat environment. The local aerodynamic heat environment wind tunnel test is meaningless if the heat flux ratio obtained by the wind tunnel test is significantly different from the actual flight situation, and the relationship and conversion relationship between the two cannot be established. Research could be carried out from two aspects on the problem of local heat environment, with combination of numerical calculation, engineering experience equation and wind tunnel test: the first is to directly make use of the universal heat flow equation to directly calculate the local thermal environment under aircraft conditions and the numerical calculation method is used for verification; the second is to fit a curve or equation about the parameters of the local thermal environment is for the specific contour of the aircraft in the complex flow area, according to a large amount of test data and calculation results, and the ground wind tunnel test data is extrapolated to the flight conditions.

2.4 Development of Flying Test

Flight test is an important part of the research on the consistency between ground and flying test. A large amount of sky flight test data is needed, especially a lot of sky tests corresponding to the ground test in the process from ground test and simulation mapping to the flying environment. Flight tests are fundamentally different from the telemetry test of general product system verification. A large number of detailed measurement data can be obtained through multiple specialized and single measurements. The data of measurement from this type of flight test provides the most lacking data in the ground and flying correlation study.

2.4.1 *Background of Flight Tests Development*

As the field of hypersonic flight technology is facing several bottlenecks related to hypersonic aerodynamics, it is urgent and necessary to develop low-cost, small-scale flight tests for the purpose of research and verification of aerodynamic related technologies. It is to research hypersonic aerodynamic force/heat related issues in this type of flight test, which makes up for the limitations of ground test simulation capabilities, and solves aerodynamic problems such as aerodynamic layout, aerodynamic force/heat accurate prediction, and efficient thermal protection in system innovation. The research results can accumulate technology for aircraft performance improvement and independent innovation, minimize the technical risk of large-scale system-level flight tests to the utmost extent, narrow the gap with hypersonic technology in developed countries such as the United States, and provide importance technology support to research on development and research and basic technologies of the follow-up related national projects.

The X-series research plan over 50 years in the United States confirms the importance of low-cost, small-scale flight tests. It can be said to be an indispensable technical means to promote the engineering of innovative technologies. Qian Xuesen attached great importance to this method of aerodynamic technology research during his lifetime, who believed that flight test, ground test technology and numerical simulation technology complemented each other, and constituted a complete means for the research on hypersonic technology.

Flight tests is to study key technical issues of single or small systems and obtain relevant data compared with system-level flight verification tests. The cost of flight test is much lower than the cost of large-scale system-level flight test due to the focus of the target, and it can greatly shorten the development cycle. Therefore, it is of great scientific and engineering significance to carry out research on flight test technology with low cost, small scale, and it will greatly promote the development of hypersonic flight-related technologies.

2.4.2 *Implementation Cases on Flight Test*

Many flight tests programs have been implemented at home and abroad over the recent years, many of which are based on basic scientific issues or individual technologies. The following is a brief introduction to such flight tests.

2.4.2.1 HIFiRE Program

HIFiRE (Hypersonic International Flight Research and Experimentation) program is a joint project of AFRL and DSTO. The purpose of this program is to develop and verify a new generation system of hypersonic aerospace, which provides basic data for the design of air and space vehicles, and provides technical support for global rapid strike and rapid entry into space.

The plan is mainly to research on the hypersonic physical phenomena and environment of hypersonic air and space flight and weapon systems, to be specific: ① Develop and verify hypersonic phenomena and calculation models; ② Guide the development of ground test technology with typical hypersonic flight environment characteristics; Develop instruments and test techniques that can be used for flight tests; ④ Develop hypersonic flight tests and load design methods; ⑤ Develop hypersonic numerical models, ground tests, and flight test correction methods.

There are many test aircraft layouts and flight methods in the HIFiRE program. The following is an example for HIFiRE-0 and HIFiRE-1.

1) HIFiRE-0

(1) Overall Situation

As a test flight platform-level flight test conducted to reduce the risk of the HIFiRE series of flight plans, HIFiRE-0 is mainly used to verify the effectiveness of flight platform systems, including verification of level sensors, accelerometers, cold push systems, flight computers, software systems and other test platform subsystems. There are more than 60% of the subsystems on the HIFiRE series of planned flight platforms that will be verified by HIFiRE-0 flight tests. HIFiRE-0 does not have any scientific problem load in order to reduce the project implementation difficulty and flight test risk.

(2) Technical Status

The contour of the HIFiRE-0 test aircraft is a simple 1067 mm long cone and a 433 mm cylinder. The cone section is made of sheet from stainless steel, and the cylinder section is made of structural aluminum. HIFiRE-0 is boosted by a two-stage Terrier-Orion solid engine to achieve a flight trajectory with a maximum flight height of over 300 km and a reentry speed of $Ma > 7.5$.

(3) Implementation Status

HIFiRE-0 completed the first flight in 2009. The risk of the HIFiRE series flight plan is effectively reduced through the HIFiRE-0 flight test, and its success rate is greatly improved. A mature and reliable test flight platform is provided for the HIFiRE program.

2) HIFiRE-1

(1) Overall Situation

HIFiRE-1 carried out flight tests based on this platform on the basis of the successful flight of HIFiRE-0, and it combined three hypersonic flow flight test contents in its mission, respectively the boundary layer transition measurement of hypersonic cone edge, boundary layer separation and aerodynamic optical effects induced by hypersonic shock boundary layer interference.

(2) Technical Status

The HIFiRE-1 test aircraft has the contour of a cone-column skirt with a cone-shaped half-cone angle of 7°, the tail skirt 33°. A large number of pulsating pressure and heat flow sensors are installed in HIFiRE-l test aircrafts, which is used for the transition measurement of conical boundary layer.

(3) Implementation Status

HIFiRE-1 completed its first flight in 2010. As a very successful low-cost flight test, HIFiRE-1 has obtained data on hypersonic boundary layer transitions, shock wave boundary layer interference, and aero-optics of more than 15G, which lays a good foundation for the HIFiRE series flight plan.

Summary to HIFiRE Test Aircraft:

The HIFiRE program is a hypersonic flight test plan developed for the development of hypersonic reentry aircrafts and hypersonic cruise aircrafts. 10 flight missions have been developed according to different mission load plans, respectively targeting different test aircraft, different trajectories and different science and technology contents. HIFiRE separately designed a flight test for the test flight platform verification in order to effectively reduce the technical risk of the flight test plan, that is, HI-FiRE-0 verified more than 60% of the subsystems on the HIFiRE flight test platform through the HIFiRE-0 flight test. The success of the test effectively ensured the smooth implementation of the HIFiRE series flight program. HIFiRE-1 carried out scientific test contents including hypersonic cone boundary layer transition, shock wave boundary interference induced separation, and aerodynamic optical effects based on the successful flight of HIFiRE-0, and obtained a large amount of basic data for hypersonic flight tests, which provides an understanding on physical phenomena and physical models for the development of hypersonic aircrafts.

It can be found from the successful implementation of the HIFiRE program that, a single key technology flight test is very necessary for complex problems in the development of hypersonic reentry vehicles and hypersonic cruise vehicles. It can effectively reduce the risk of development, shorten the development period, and provide necessary scientific understanding and technical reserve for engineering.

2.4.2.2 SHEFEX Test Aircrafts

SHEFEX (SHarp Edge Flight Experiment) is a flight test program by DLR to carry out research on new concept spaceflight aerodynamic layout, sharpened leading edge and combined surface layout. In addition, the SHEFEX series of flight tests can obtain important knowledge in hypersonic aero-physics, which becomes a test platform for new concept and new technology flight test verification. SHEFEX program is to develop aerospace technologies in the future such as reentry vehicles, hypersonic aircraft and TSTO.

The preliminary plan of SHEFEX consists of three flight test projects, respectively that, SHEFEX-I completed the flight test in 2005, SHEFEX-II plans to conduct flight test in the near future, and SHEFEX-M implemented the flight test in 2015.

1) SHEFEX-I

(1) Overall Situation

The main objective of the SHEFEX-I flight test is to verify the impact of the sharpened leading edge on the aerodynamic performance of the hypersonic aircraft faced with the advanced hypersonic transportation system, and to evaluate the performance of the composite surface (TPS). The hypersonic physical model, calculation method and ground test data developed by DLR were verified through flight test data at the same time.

(2) Technical Status

The flight system consists of the first-level engine Brazilian S30, the second-level engine improved Orion, the SHEFEX-I test aircraft and the fairing. The flight control is completed through test aircraft without flight control device-level rocket.

(3) Implementation Status

SHEFEX-I completed the flight test successfully in 2005, with a flight time of 550 s, an altitude of 300 km, a range of 211 km, a test window of 90–20 km for the reentry section, and a 40 s. It obtained a large number of hypersonic reentry aerodynamic/thermal data in the flight test. At the same time, it also verified the key technologies such as flight control system and hypersonic flight test system, and provided a mature test flight platform for the follow-up flight test of SHEFEX.

2) SHEFEX-H

(1) Overall Situation

The SHEFEX-H test aircraft is the second step in the route map of the SHEFEX series of development projects, of which the flight test window and maximum flight Mach number are increased from 20 s, $Ma = 6$ of SHEFEX-I to 45 s, $Ma = 11$ respectively. Its main objectives are: to test hypersonic active pneumatic control systems (emphasis on guidance and navigation control), new inertial navigation platforms, star sensors, verification of active cooling heat protection, sensors (temperature, heat flow and pressure), high temperature resistant antenna technology, etc.

(2) Technical Status

The consist of the test aircraft is basically the same as that of SHEFEX-I, with a total length of 12.6 m and a mass of 6.7t.

The motor of solid rocket serves as the first stage propulsion device. The test window starts at a height of 100 km in the re-entry section, at which height the atmospheric effects begin to appear. The aircraft will fly at the maximum speed of $Ma = 11$ for *45 s* in the reentry section in the current flight envelope.

(3) Implementation Status

SHEFEX-H is originally planned to make its first flight in 2010, but it was delayed later due to the launching system and loading system and now it is planned to launch in the near future.

Summary to SHEFEX Test Aircraft:

The program of SHEFEX is a test flight plan developed for research on aerospace technology such as new concept reentry vehicles, hypersonic aircraft, TSTO, etc. DLR has formulated a step-by-step development plan for gradual execution in order to achieve the series of research targets. Firstly, SHEFEX-I flight test was conducted to verify the key technologies of hypersonic reentry, such as aerodynamic layout and heat protection, test flight platform technology for launch and boost. At the same time, the understanding of hypersonic physical phenomena and physical models is obtained through the flight test, to verify and improve the theoretical calculation model of hypersonic aerodynamic heat environment and ground wind tunnel test technology. With this basis, SHEFEX-H performs the key technologies for suborbital reconstruction such as navigation control, active cooling, heat protection, sensors (temperature, heat flow pressure), etc.

It can be found from the implementation of the SHEFEX research program that only through flight tests can it obtain sufficient knowledge of the hypersonic physical phenomena and physical models for the new concept hypersonic reentry aircraft, so as to achieve verification and calibration of the results of the calculation model and ground wind tunnel test. The maturity of key technologies such as aerodynamic force, heat protection, etc. must be relied on the flight test for verification. It is essential for the flight test verification of a single technology in order to reduce the test risk.

2.4.2.3 HyBoLT

1) Overall Situation

HyBoLT (Hypersonic Boundary Layer Transition) is a research project on hypersonic aerodynamic basic issues supported by NASA, mainly focusing on issues of hypersonic boundary layer transition and transition flow control. The issues of hypersonic boundary layer transition and transition flow control are the precondition and basis for the design of heat protection and reduction of heat and drag on the aircrafts. The current theoretical analysis technology lacks experiments data calibration for the transition of the boundary layer after the spacecraft and other orbiting vehicles enter the atmosphere, and ground wind tunnel test technology cannot simulate the actual flight state, therefore NASA carried out research on single flight test on hypersonic boundary layer transition measurement.

2) Technical Status

The flight system of HyBoLT consists of a multi-stage booster rocket ALV X1, an adapter, a command module, and a HyBoLT aircraft. The tower launcher with no recovery device has a pointed wedge shape with a half wedge angle of 6°, a leading edge with radius of 3.8 mm, a wedge length of 2.28 m, and the side is conical. The flight trajectory of HyBoLT is parabolic, and its test window is the first 65 s.

3) Implementation Status

It conducted a flight test for the project in 2008. The flight mission was terminated on 20 s after launch due to the abnormality of the booster rocket control system, but a large amount of boundary layer transition flight test data has been obtained.

4) Summary

HyBoLT is a program on single scientific mission test carried out by NASA to solve the super basic aerodynamic problems encountered in the development of hypersonic reentry vehicles and hypersonic cruise vehicles and the research objects are sharp wedge boundary layer transition and transition flow control. A large amount of boundary layer transition test data was obtained in just 20 s although the HyBoLT program ended with the failure of the first flight, which provided technical support for the design of hypersonic aircraft.

2.4.3 Direction of Flight Tests Development

The aerospace aerodynamic flight test technology will be established through the flight test to realize the three sets of aerospace aerodynamic research theory, ground test and flight test. The aerodynamic measurement technology for flight tests can be applied to research on new concepts of key aerodynamic technologies and important aerodynamic fundamental issues for the development of hypersonic vehicles such as shock wave/boundary layer interference, hypersonic boundary layer transition, and real gas effects in hypersonic aircrafts, which provides flight test data verification for theoretical numerical model and ground wind tunnel test measurements, and the obtained flight test results can be used directly in the research on the consistency of the ground and flying. It provides an important experimental basis in order to establish the relevant laws between ground and flying tests.

The flight test will become a practical fine flying test, an extension of the ground test, and a data input and test verification of the consistency between the ground and flying tests.

2.5 Summary and Outlook

In general, there is a strong experience and subjectivity for the traditional analysis method on the consistency of the ground and flying test.

The results obtained are local data dependent in a lack universality. However, the technological reserves for new types of aircrafts are insufficient, and the accuracy of prediction is higher than that of traditional aircrafts. The research on the accurate prediction of the consistency of aerodynamic heat between ground and flying is the direction for the breakthrough of new aircrafts development.

The relationship between ground and flying test is quite different, as the design and verification methods for aircraft test are imperfect, the research on basic theory and objective laws such as the relationship between ground and flying are inadequate, and there is a lack of comprehensive and accurate predictions on the real environment. There is still a great demand for accurate prediction on the flight environment with the test database and verification on the consistency of the ground and flying, although a large amount of ground and flight test data has been accumulated. There are generally problems as ground testing is limited by ground conditions, such as inaccurate mechanical environments, intrinsically different ground and flying test conditions, and test measurement techniques that cannot meet the requirements on accurate measurements. The characteristics of ground tests itself can easily lead to over-test or short-test. It has always been an important subject of aerodynamic thermal research on how to improve the effectiveness of the ground test and get closer to the flying environment.

There are three aspects included for the research on the verification of aerodynamic heat consistency between ground and flying: the first is to convert the measurement data of different wind tunnel test conditions to each other, verify the correlation method, adopt the established correlation conversion method, and consider the model scale, Reynolds number, Mach number, total temperature, total pressure, wall temperature ratio and other parameters, so that the correlation, conversion and prediction of test results can effectively verify the ground and flying conversion method. The Mach number, Reynolds number, total temperature, total pressure, etc. of the test flow field can better simulate the flight trajectory parameters of the test vehicle if the wind tunnel test adopts a 1: 1 model, then the heat flow and dimensionless heat flow of the test results will be the same with flight conditions, which can provide a basis for the verification of the ground and flying conversion method. The second is to verify with theoretical calculation methods. The typical flight state and the aerodynamic heat along the trajectory are predicted for a typical shape aircraft, through the conversion of the wind tunnel test data. At the same time, the model with different reduction ratios under the wind tunnel test conditions, theoretical calculation results under different flow conditions can be obtained with numerical and engineering calculation methods. The method of ground and flying tests conversion is verified through the comparison between the prediction results of the ground and flying conversion and the prediction results of the theoretical calculation. The third is to verify the method with typical flight results at home and abroad. It is to extrapolate to the flight conditions according to the wind tunnel test data, predict the flight results, and verify and improve the correlation method by comparing with the typical flight test results abroad and the typical flight test results that have been already carried out in China.

Finally, it is necessary to evaluate the accuracy of the method for conversion of ground and flying aerodynamic heat and give the confidence interval of the extrapolated data. It is planned to carry out a theoretical analysis on the similar correlation parameters of different boundary layers to research on the uncertainty generated during the derivation of the correlation parameters of the ground and flying conversion, and comprehensively research on the ground and flying conversion method and the uncertainty of the hole test data with combination of the error and uncertainty analysis of the ground wind tunnel test data, so as to establish an uncertainty assessment method for the prediction results of the ground and flying conversion. The aerodynamic heat under different conditions are calculated and converted into ground and flying respectively according to the typical wind tunnel test conditions and flight conditions. The error zone caused by the world and earth conversion is determined with comparison on the calculation results with the ground and flying conversion results.

A relatively complete system from wind tunnel test to flight correlation correction is established based on the results and experience of relevant research at home and abroad, to research on the factors and their changing laws between the difference between wind tunnel test and real flight to form a set of methods for practical and reliable wind hole test data correction and use. A universal design specification, and the law on correlation of aerodynamic parameter characteristics can be given based on several flight test data finally.

The environment of the new hypersonic aircrafts will become more and more complex in the future, and the current technology on ground and flying consistency verification cannot meet the reliability and other requirements on evaluation. It is needed to carry out in-depth research on prediction and verification on the consistency of the ground and flying tests, in which it can predict the increasingly complex aerodynamic, heat, load, electromagnetic and other flight environments faced by the aircraft through ground tests or virtual simulations, and simulate the flight environment and verification of various system tests on the ground, so as to support aircraft to accurately figure out the actual flight boundary and its own capability boundary.

References

1. Praharaj SC, Roger RP, Chan SC et al (1997) CFD computations to scale jet interaction effects from tunnel to flight. AIAA Paper, 97-0406
2. Ewald BFR (ed) (1998) Wind tunnel wall corrections, vol AGARD-AG-336. AGARD
3. Viswanath PR, Rajendra G (1990) Sting corrections to zero—lift drag of axisymmetric bodies in transonic flow 13. Aeronaut J 94:279–288
4. Cahn MS (1956) An experimental investigation of sting—support effects on drag and a comparison with jet effects at transonic speeds. NASA report
5. Steinle F, Stanewsky E (1982) Wind tunnel flow quality and data accuracy requirements. Advisory group for aerospace research and development neuilly—sur—seine (France)
6. Ni Z, He D (2000) Research and application of correction method of equivalent dynamic pressure tunnel wall interference. J Aerodyn 18(1):86–91
7. Yin L, He Z, Yu Z et al (2000) Research on correction method for the wall interference of subsonic high angle of attack model test. Exp Fluid Mech 03:37–41
8. Zhang R, Wang X, Li Z et al (2006) Research on the Interference of tail boom support sting in low-speed wind tunnel]. Exp Fluid Mech 20(3):33–38
9. Chen D, Zhao X, Wu K et al (2002) Research on wind tunnel and flight correlation of nonlinear aerodynamic characteristics of small aspect ratio aircraft. J Aerodyn 20(1):72–77
10. Yan C, Yu J, Li J (2006) Research on some issues of format and reseau effect in heat flow CFD calculation. J Aerodyn 24(1):125130
11. Pan S, Feng D, Ding G et al (2010) Reseau correlation and convergence in aerodynamic thermal simulation. J Aeronaut 31(3):493499
12. Zhang X, Wu Y, Wang J (2012) Numerical simulation research on aerodynamic characteristics of axisymmetric reentry module. J Appl Mech 29
13. Cai G, Xu D (2012) Hypersonic aircraft technology. Science Press, Beijing
14. Tang D (2015) Boundary layer transition. Science Press, Beijing
15. Kirsten PW, Richardson DF, Jacob D (1983) Ground/flight test techniques and correlation. AGARD CP—339

Chapter 3
Bluff Body Overflow and Drag Reduction Control

3.1 Flow Over Blunt Body

As a non-streamline body, bluff body is widely applied in many engineering projects such as aerospace, construction, transportation, and environment, which has a larger and even more overwhelming differential pressure resistance compared to streamlines in case of aircraft, building structures, bridges, automobiles, heat exchangers, eddy-current flowmeter, etc., and is a classic branch in the research field of fluid mechanics. There are rich passing flow structures around the bluff body. For example, the surface of the bluff body will generate a wide range of boundary layers of separation and form a wide wake flow with vortex shedding.

3.1.1 Separation of Boundary Layer

A thin boundary layer will be formed on the surface of the object under the action of the fluid viscosity when the fluid flows around the object. There is a large velocity gradient distribution in the boundary layer along the normal direction of the object surface, and it will also show a large non-negligible viscous force, even if the fluid is very viscous. It will form a adverse pressure gradient on the surface of the object due to the viscosity of the fluid itself, and then the boundary layer separation (flow separation) takes place. Take separation when fluid flows around the curved wall as an example, as shown in Fig. 3.1.

The physical interpretation to the separation of boundary layer is that the intensity of pressure of the mainstream can act on the object surface through the boundary layer. The forward pressure gradient is sufficient to overcome the effect of viscosity in the AC forward pressure gradient and keep the fluid particles moving at a certain momentum. The pressure in front is large in the CE reverse pressure section, and

M. Zhao, *Prediction and Validation Technologies of Aerodynamic Force and Heat for Hypersonic Vehicle Design*, Springer Aerospace Technology,
https://doi.org/10.1007/978-981-33-6526-1_3

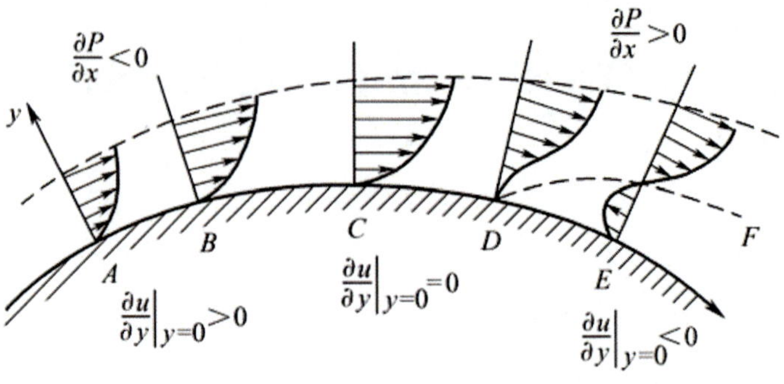

Fig. 3.1 Separation of the boundary layer

the moving fluid particles must overcome the viscous effect and resist the pressure difference, so that the moving fluid particles gradually decelerate under the double viscosity and adverse pressure until the kinetic energy is exhausted to point D. However, the fluid particles in the DE section will stagnate or even move backwards under the effect of adverse pressure, which causes a back flow. The back flow squeezes the fluid particles in the boundary layer outside the mainstream, so that the boundary layer is separated from the wall surface for the separation of the boundary layer.

In mathematics, the analysis can be carried out with boundary layer theory proposed by Prandtl. The Prandtl equation of boundary layer is applied to analyze the effect on the velocity distribution from the pressure gradient on the wall. The component equation along the x direction of the curved wall coordinate is:

$$\rho\left(u\frac{\partial u}{\partial x}+v\frac{\partial u}{\partial y}\right)=-\frac{dp}{dx}+\mu\frac{\partial^2 u}{\partial y^2} \tag{3.1}$$

The position at wall surface meets u = 0, v = 0, which can be obtained from Eq. (3.1)

$$\frac{\mathrm{dp}}{\mathrm{dx}}=\mu\frac{\mathrm{d}^2\mathrm{u}}{\mathrm{dy}^2} \tag{3.2}$$

Equation (3.2) shows that the second derivative of the velocity profile in the normal direction of the wall is the same with the sign of the pressure gradient in the x direction. There is $\frac{\partial^2 u}{\partial y^2}<0$ in the forward pressure gradient region (AC segment), the velocity profile convex can be obtained from the differential nature of the function. There is $\frac{\partial^2 u}{\partial y^2}=0$ in position C. And there is $\frac{\partial^2 u}{\partial y^2}>0$ in the reverse pressure gradient area (CE section). The velocity profile is concave and the curvature along the moving direction of the flow gradually increases to point D, $\frac{\partial u}{\partial y}=0$. The velocity profile is tangent to the y direction. The velocity profile continues to be concave after passing point D, the velocity becomes a minus value, and a reverse flow takes place. The DF line is called the short surface academically, and the part behind the DF line is the separation area.

It can be seen from the above analysis that the root cause of boundary layer separation is the existence of fluid viscosity, and the condition for the separation is the existence of adverse pressure gradient. The actual separation is caused by the stagnation and back flow of fluid particles.

It is shown in Fig. 3.2 that the convex wall blunt body develops from the initial boundary layer of motion starting from static, photographed by Prandtl in 1943. The adverse pressure gradient is very small, and the flow field is close to the inviscid flow (Fig. 3.2a) when the object is just started. The adverse pressure gradient increases at the rear as the object begins to accelerate, and a separation vortex appears near the rear stagnation point (Fig. 3.2b), after which, the separation point moves upstream (Fig. 3.2c) and finally, the separation vortex strengthens into a circular vortex (Fig. 3.2d).

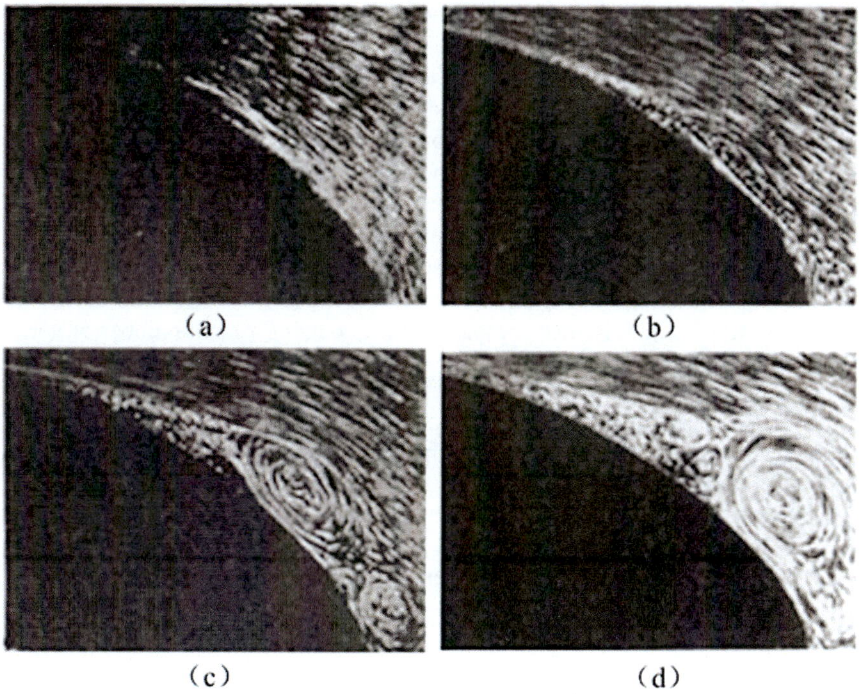

Fig. 3.2 Separation of boundary layer on convex surface

3.1.2 Flow Resistance Around Bluff Body

It can be known from the previous analysis that it will be affected by the viscosity of the fluid when an object is immersed in a viscous fluid.

The frictional resistance caused by the shearing stress is also stressed by the pressure difference resistance caused by the flow separation. The pressure resistance is also called shape resistance since the magnitude of the adverse pressure gradient is related to the shape of the object surface.

The frictional resistance is caused by the viscous shearing stress of the fluid on the object surface. The viscous shearing stress is proportional to the fluid viscosity coefficient and velocity gradient, according to Newton's law of viscosity. The viscous shearing stress produced by different velocity distributions on the object surface is different for the same fluid. The flow state in the boundary layer of the object surface is related to the friction resistance when the fluid flows around the object with a large Re flow, of which the value can be solved through the boundary layer theory, and the friction resistance of the turbulent boundary layer is larger than that of the laminar boundary layer.

The pressure drag is the resistance caused by the combined pressure of the fluid acting on the object surface. The boundary layer is likely to separate into a vortex on the leeside when the fluid flows around the bluff body, and the pressure at the hind side cannot be restored to be symmetrical to the front. Therefore, there will be differential pressure resistance in front and back of the object. The more severe the separation at the hind side, the greater the differential pressure resistance.

The theoretical analysis on the pressure drag on the passing flow object is more difficult than the analysis on frictional resistance. The main barrier is that the flow field in the wake flow area is very complex, which contains a large number of vortexes and velocity fluctuations. These vortexes and pulsations will also quickly trigger turbulence even when the Reynolds number around the flow is not large, which makes analytical analysis difficult to carry out. Therefore, most of the resistance on contour comes from test measurements. There are two indirect measurement methods commonly used in addition to using a mechanical sensor to directly measure the force component of the flow direction: One of the workable methods is to measure the velocity distribution on the section farther downstream from the flow object in the wind tunnel, and then calculate the resultant force of the object with the momentum theorem, representing the contour resistance F_{Df} with the resultant force, as it is shown in Fig. 3.3.

$$F_{Df} = \int_{-\infty}^{+\infty} U(U_{\infty} - U)dy + \rho \int_{-\infty}^{+\infty} [\langle v'^2 \rangle - \langle u'^2 \rangle]dy \tag{3.3}$$

In the equation: U_∞ is the velocity of the incoming flow. $[\langle v'^2 \rangle - \langle u'^2 \rangle]$ is the stress term in the normal direction and the flow direction. Another method is to set up a pressure measuring hole on the surface of the object, measure the pressure distribution, and divide the pressure along the object area to estimate the differential

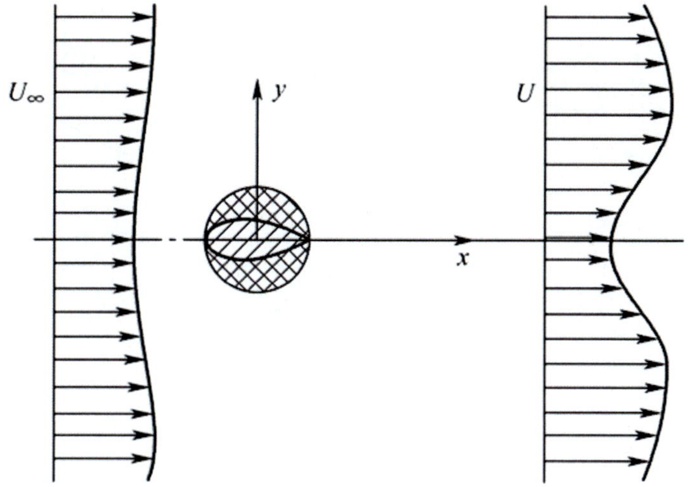

Fig. 3.3 Symbol explanation for resistance calculation with momentum method

pressure resistance. The pressure drag is usually expressed by the non-dimensional pressure drag coefficient C_{Df} in engineering:

$$C_{Df} = \frac{F_{Df}}{\frac{1}{2}\rho U_\infty^2 A} \tag{3.4}$$

In this equation: A is the reference area, which is generally valued by the projection area perpendicular to the incoming flow. $\frac{1}{2}\rho U_\infty^2$ is the incoming flow.

3.1.2.1 Flow Field and Resistance of the Flow around Cylinder

As a typical bluff body, cylinder is a classic model for analysis on the essential issue of the flow around the bluff body. It has attracted the attention of researchers in the past 50 years of research.

1. Flow Field Characteristics of the Flow around a Cylinder

It is found from experimental research that the flow structure on the surface of the cylinder will vary with the change of Reynolds number. Figure 3.4 is a schematic diagram of the flow field structures of the cylinder under different Reynolds numbers.

(1) The range is called flow at low Reynolds number or creeping flow when $Re < 5$. This kind of flow is extremely slow, and there is no vortex outflow, that is, the boundary layer is relatively stable, and the tail vortex regularly merge and cancel each other out;

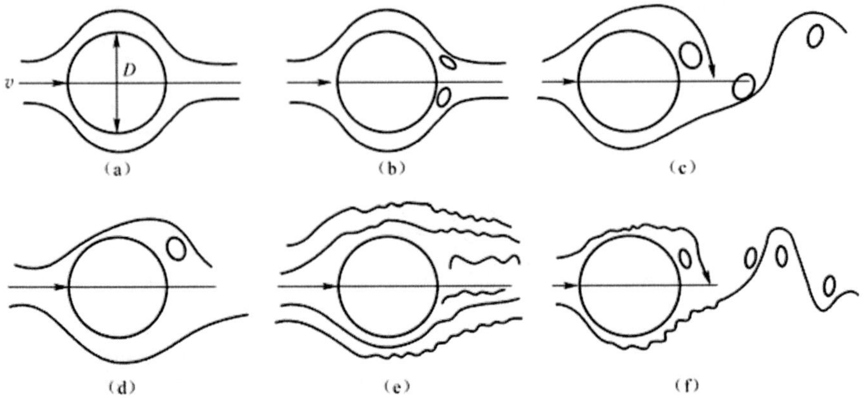

Fig. 3.4 Changes of cylindrical flow field structure with Reynolds number. **a** $Re < 5$; **b** (5–15) < $Re \leq 40$; **c** $40 < Re \leq 150$; **d** $150 < Re < 300$; **e** $3.0 \times 10^5 < Re < 3.5 \times 10^6$; **f** $Re \geq 3.5 \times 10^6$

(2) As the flow velocity increases when $5 < Re \leq 40$, the gradually unstable boundary layer starts to flow away from the rear of the cylinder, which forms into two symmetrically stable vortexes;
(3) Cross vortex trains of laminar flow forms behind the cylinder when $40 < Re \leq 150$;
(4) The laminar vortex train begins to transit to turbulent flow when $150 < Re < 300$, so that the boundary layer also begins to transition to turbulent flow.
(5) The boundary layer is all in a turbulent state when $3.0 \times 10^5 < Re < 3.5 \times 10^6$, and the wake becomes narrow and chaotic.
(6) The wake becomes a cross vortex when $Re \geq 3.5 \times 10^6$.
(7) The left and right sides of the wake of the object will produce symmetrical vortexes that are paired, alternately arranged, in opposite directions of rotation when such a fluid flows around a solid, which is called Karman Vortex Street.

The Czech scientist Strouhal (1878) put forward the empirical equation of the cylindrical vortex release frequency through a research on a tweet on the wire in the wind in the nineteenth century:

$$S = \frac{f_D}{U_\infty} = 0.198\left(1 - \frac{19.7}{Re}\right) \tag{3.5}$$

In this equation: S is the Strouhal number and D is the diameter of the cylinder. $Re = \frac{\rho U_\infty D}{\mu}$ indicates that S is affected by Re.

Many scholars have found through experiments in later stage that, there can be a regular Karman vortex street behind the cylinder in the range of $Re = 60$–5000 and the S number remains almost maintains constant in the range of $Re = 300$–2×10^5. And then the vortex street is no longer regular until the vortex street reappeared when $Re \geq 3.5 \times 10^6$. The relationship between the cylinder S and the Reynolds number was experimentally measured, and its value was stable at $S = 0.20$ when $Re = 1000–2 \times 10^5$.

Strouhal proposed the empirical Eq. (3.5) on the basis of ignoring the structure of the flow field. Ro-Shko [1] proposed the wake Strouhal number S* in 1955 based on the structure of the flow field:

$$S^* = \frac{fL^*}{U^*} \tag{3.6}$$

The characteristic length D of the bluff body is replaced by width L^* formed by trailing vortex, while the speed U is replaced by the speed U^* formed by trailing vortex. The S^* value is stable at about 0.16 for the distribution curve of the Strouhal number defined in different ways, compared with the large fluctuation of S [2, 3].

Williamson and Brown [4], has improved the empirical equation on the basis of the wake Strouhal number (S^*), on account of a certain relationship of the vortex frequency and the thickness of the boundary layer of the object surface with the transport speed of the separated shear layer.

$$S^* = \left(\frac{f_D}{U_\infty}\right)\left(\frac{U_\infty}{U_S}\right)\left(1 + \frac{2\delta}{D}\right) \tag{3.7}$$

The width L^* is formed with $D + 28$ replacing the tail vortex, where δ is the thickness of the boundary layer. The velocity at the shear layer separation point is replaced with U_S.

Among them, the velocity U_s of the shear layer and the cylindrical back pressure coefficient $C_{Pb} = (2P_b - P_s)/pU^2{}_\infty$ (P_b is the total pressure of the stagnation point of the cylinder, P_s is the static pressure value of the free flow), there is the following relationship [1]:

$$\frac{U_s}{U_\infty} = \sqrt{1 - C_{P_b}} \tag{3.8}$$

Therefore, there are the following conversion equations for the universal Strouhal number S and the trailing Strouhal number S^*:

$$S = S^* \sqrt{1 - C_{P_b}} (1 + \frac{2\delta}{D})^{-1} \tag{3.9}$$

The improved Strouhal number S and the distribution of $\sqrt{1 - C_{P_b}}(1 + \frac{2\delta}{D})^{-1}$ is concentrated on a line with a slope of 0.176, that is, $S^* = 0.176$. Compared with S^* with $Re(55 \sim 1.4 \times 10^5)$, the improved S^* distribution is obviously concentrated and stable at 0.176.

Combined with the structure of its flow field, it can be divided into the following areas: the fluid flowing on the surface of the cylinder leaves the wall surface near the apex (90°) of the cylinder, and the separated fluid is formed into a pair of fixed symmetrical vortex (attached vortex) in the downstream of the cylinder.

Regarding the relationship between Strouhal number S and Re, Williamson and Brown [4] delivered a more detailed fitting relationship based on the characteristics of S-Re distribution in the test results, who proposed

$$S = \left(A + \frac{B}{\sqrt{Re \frac{C}{\sqrt{Re}}}} \right) \tag{3.10}$$

The description of the flow field around a cylinder according to different Reynolds number ranges based on the proportional relationship of the thickness of the boundary layer ∂ to $1/Re$. Another key parameter is the cylindrical back pressure coefficient C_{Pb} in addition to the aforementioned Strouhal number. -C_{Pb} is often used in the corresponding researches instead of the coefficient of back pressure C_{Pb} to describe its effect on the cylindrical wake [5, 6], which directly affects the flow field structure in the cylindrical wake area. The distribution diagram of -C_{Pb} under a smooth cylinder with Reynolds number Re is measured through experiment [7].

1. -C_{Pb} remains stable when $Re < 49$ and the boundary layer in this area is in a laminar state, the fluid in the vortex forms a closed circuit which becomes a "Dead Water Zone".

(1) When $Re = 10$, there is a pair of standing vortexes at the rear of the cylinder, the Reynolds number Re gradually rises up, the dead water zone gradually stretches, and the asymmetry of the flow field in front and back of the cylinder gradually becomes more obvious. This range of Re is called the symmetric wake flow zone.

(2) -C_{Pb} showed an increase when $49 < Re < 190$ (A-B), and the backflow zone behind the cylinder became unstable. The backflow zone showed strong velocity fluctuations with the increase of Re, which spread upstream and gradually disintegrated with adherence to the vortex. The flow field showed an unsteady state, and the vortex on both the upper and lower sides of the trailing edge of the cylinder periodically fell off, forming a "Karman Vortex Street", which was a laminar vortex train.

(3) -C_{Pb} showed a tendency to slowly decrease and then increase, and reached a peak at Re = 260 when $190 < Re < 260$ (B-C). The instability of the cylindrical backflow zone further increased, and the laminar vortex transits to turbulent vortex. The flow field shows a certain three-dimensionality, Mode A.

(4) -C_{Pb} showed a downward trend when $260 < Re < 1000$ (C-D), and the three-dimensional effect of the flow field further increased, Mode B. There is a small-scale three-dimensional disturbance appeared, and its tail vortex formation length became larger.

(5) -C_{Pb} enters the ascending phase again when $1000 < \mathrm{Re} < 2 \times 10^5$ (D-E). It is shown in experiments that, the boundary layer changes from laminar to turbulent in this area, the Reynolds stress on the surface of the cylinder increases, and the length of its tail vortex formation becomes smaller.

(6) The E–F-G area, is also known as the transition zone, in which the resistance coefficients of -C_{Pb} and the cylinder both decrease rapidly. The boundary layer on the surface of the cylinder changes from laminar to turbulent at this time, and the boundary layer separates and reattaches, but presents an asymmetric state [8]. The separation point moves backward (140°) and the separation area gradually shrinks. The shedding of vortex is irregular, and the overflow frequency of vortex is random at a wide frequency.
(7) The GH area is called the supercritical area, in which it shows a symmetric separation and reattachment phenomenon in the boundary layer. The separation point moves forward again in this area, it re-establishes a more regular quasi-periodic vortex street in the wake flow, and -C_{Pb} rebounds.
(8) The boundary layer is completely transformed into turbulent flow in the H-J area, and regular shedding of vortex no longer occurs.

2. Resistance characteristics of flow over a cylinder

It is clear through the previous analysis of the flow field around the cylinder that the resistance of the flow around the cylinder is directly affected by the structure of the flow field. The resistance characteristics of the cylinder has been introduced below.

Distribution of Pressure Coefficient on the Surface

The pressure at the rear changes dynamically with the generation and shedding of the vortex when the actual fluid flows around the cylinder due to the flow separation at the rear. There is much difference between the pressure distribution on the rear surface of the cylinder and the inviscid flow. The pressure cannot be restored to the same level as the front, and mostly remains negative (surface pressure).

The pressure in the back part cannot be restored to the level of the front part, and the resultant force of pressure is obtained from integration on the area of the cylinder, which is the pressure drag.

The distribution curve of the pressure coefficient on the surface of the cylinder is measured by the test when $Re = 1.4 \times 10^5$. The solid point is the average pressure coefficient, and the vertical thin line is the fluctuation range of the pressure coefficient at different positions. The pulsation frequency of the pressure coefficient is consistent with the vortex off frequency behind the cylinder, which also further proves that the pressure differential resistance experienced by the cylinder is directly related to the flow structure on its surface.

The Changes of Cylinder Resistance Coefficient with Reynolds Number

The relationship curve of C_D-Re is measured by the two-dimensional smooth cylinder test. The relationship between the speed and the characteristics of the change of the drag coefficient can divide the curve into 6 areas according to resistance, and the division of these 6 areas is similar to some contents to that of the -C_P-Re curve. It is noteworthy that, its value is stable at $Cd = 1.2$ at $Re = 1000$–2×10^5. There is an area where resistance drops rapidly at $Re - 2 \times 10^5$, and this phenomenon is called "Resistance Crisis".

(1) $Re << 1$, is called low Reynolds number flow or creeping flow. There is almost no flow separation, and the resistance is mainly frictional resistance, which is proportional to the speed to the first power.
(2) There is flow separation when $1 \leq Re \leq 500$. There is a pair of standing vortexes at the rear of the cylinder when $Re = 10$. Vortexes are alternately released from the rear of the cylinder to form a Karman Vortex Street when $Re > 100$. The resistance is composed of friction resistance and differential pressure resistance and is roughly proportional to the speed to 1.5th power.
(3) The flow separation is serious when $500 \leq Re < 2 \times 10^5$. Starting from $Re =$ 104, the boundary layer starts to separate even from the front of the cylinder, and the vortex street breaks into turbulent flow, forming a wide separation zone. The resistance is dominated by differential pressure resistance and is proportional to the speed to the second power, that is, C_d hardly changes with the Re number.
(4) The laminar boundary layer becomes a turbulent boundary layer when $2 \times 10^5 \leq Re \leq 5 \times 10^5$, then the separation point moves backward, the resistance decreases, and the C_D falls down. $C_D = 0.3$ reaches the minimum when $Re = 5 \times 10^5$, and the separation zone is the smallest at this time.
(5) The separation point moves forward again when $5 \times 10^5 \leq Re \leq 3 \times 10^6$, and C_D rises.
(6) C_D has nothing to do with Re, and is called the self-mode zone.

3.1.2.2 Resistance Coefficient of Objects in Different Contours

1. Two-dimensional Bluff Body

The resistance coefficient CD of typical two-dimensional bluff body and Strouhal number S are listed in Table 3.1.

The picture is the schematic diagram of the cross-section over the flow, in which, the Reynolds number of the passing flow $Re >$ 104, and the characteristic dimension of the Reynolds number is the height of the cross-section. It can be seen from the table that the drag coefficient is not only large but almost similar to $C_D = 2.0–2.3$ for all sections with steep end faces and sharp corners. This is because the fluids are directly separated from the corner points, and the contour of the wake flow zone is almost the same. It has been proved in the tests that the drag coefficient of such objects hardly changes with Reynolds number, but the drag coefficient can be reduced by 25–50% after adjusting their azimuth. It can reduce the shape resistance to raise the aspect ratio of the cross section. For example, the resistance decreases by nearly 40% if the aspect ratio of the rectangular column increases from 1 to 4 [9].

2. Three-dimensional Bluff Body

The drag coefficients of typical three-dimensional bluff bodies are listed in Table 3.2. The picture is a three-dimensional schematic diagram, in which, the attack area is taken as the characteristic area, Re $\geq$ 104. As it can be seen from the table, the

Table 3.1 Two-dimensional bluff body resistance coefficient [7, 9] (infinite length, A is the attack area, $Re \geq 10^4$)

Two-dimensional object	C_D	S	Two-dimensional object	C_D	S
Cylinder	1.2	0.21	Square column	2.1	0.13
Semi-cylinder	1.2	0.21	Square column (angle of attack 45°)	1.6	0.15
	1.7	0.16	Rectangular (2:1)	1.8	0.08
Flat plate	2.0	0.16	Rectangular (1:2)	2.4	0.14
Semicircular Tube	1.2	0.21	Wedge	1.6	0.18
	2.3	0.13		2.0	0.13

Table 3.2 Drag coefficient of three-dimensional bluff body [7, 9] (A is the attack area, $Re > 10^4$)

Three-dimensional object	C_D	S	Three-dimensional object	C_D	S
Sphere	0.47	0.19	60° cone	0.80	0.19
Hemisphere	0.4	—	90° cone	1.2	0.16
Hemisphere	1.2	—	Disk	1.1	0.13

drag coefficient differs from 2.8 to 3.5 times when compared with back-to-back flow when the planes of the hemisphere are directly opposed to the flow. It is noteworthy that, the Strouhal number S decreases as the resistance increases CD regardless of whether it is two-dimensional or three-dimensional [10].

3. Sphere

A sphere can be used as a typical example for a three-dimensional bluff body. The resistance curve of a sphere is shown in Fig. 3.5. $Re = pUd/\mu$ where U is the incoming velocity, p is the density of the fluid medium, and d is the diameter of the sphere and μ is the viscosity coefficient. It is called low Reynolds number flow when $Re < < 1$, and there is no separation for the fluid flows over the sphere, which can be described by the N-S equation in the form of spherical coordinates, however, the

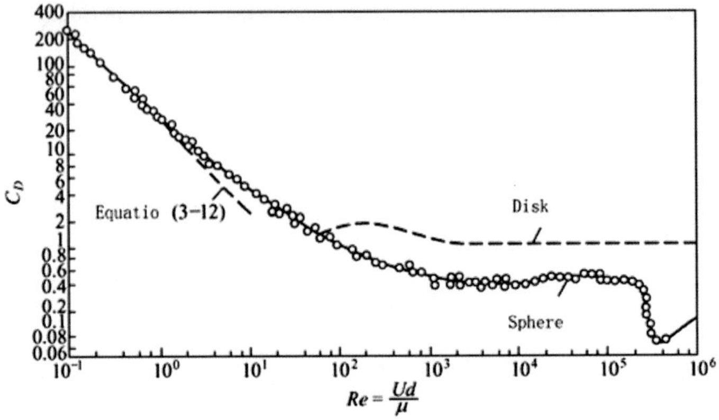

Fig. 3.5 Drag coefficient of sphere [9]

N-S equation can be greatly simplified since the inertial force is much smaller than the viscous force. The NS equation can be solved to obtain the distribution of shear stress and normal stress on the spherical surface with using of the non-slip condition of the object surface and the infinite current flow condition, and then the analytic expression of the total resistance of the sphere can be obtained through the integration on the components of the two stresses in the direction of flow over the spherical area, which is also known as the Stokes equation for ball resistance:

$$F_D = 3\pi \mu d U \tag{3.11}$$

It is indicated in the calculations that there are both frictional resistance (accounting for 2/3) and differential pressure resistance (accounting for 1/3) in the total resistance. The presence of the latter indicates that the pressure distribution on the front and back hemispheres is also asymmetric. Then the drag coefficient of the flow over the ball with low Reynolds number can be obtained as the below:

$$C_D = \frac{3\pi \mu d U}{\frac{1}{2}\rho U^2 \frac{\pi}{4} d^2} = \frac{24}{\rho U d/\mu} = \frac{24}{Re} \tag{3.12}$$

It is shown in the test that the above theoretical results are in agreement with the measured data.

Similar to the flow over a cylinder, the flow over a sphere starts from $Re > 1$ to separate at the rear, and the pressure drag increases accordingly. The proportion of viscous resistance in the total resistance keeps declining as the Reynolds number increases, and it only accounts for 5% of the total resistance until $Re = 1000$. The drag coefficient remains almost the same in the range of $103 < Re < 3 \times 10^5$, but lower than the cylinder with the same diameter ($C_D = 1.2$) or even smaller ($C_D = 0.4$). The drag coefficient also suddenly fell from 0.4 to 0.1 when it is to $Re = 3 \times 10^5$. Prandtl conducted an experiment before, in which a wire ring was placed in front of the smooth ball to artificially convert the laminar boundary layer into a turbulent boundary layer in advance. As a result, the separation point moved from the original $\theta = 80°$ to $\theta = 120°$ around, and the drag coefficient drops obviously. This is because the velocity profile in the turbulent boundary layer is full and overcomes on the separation ability enhanced than laminar flow, which is also the mechanism of many subsequent drag reduction controls.

3.2 Drag Reduction for Blunt Body

According to the analysis on the flow and resistance characteristics of a typical bluff body in Sect. 3.1, the source of resistance to the bluff body is mainly the frictional resistance caused by the viscosity of the fluid when the Reynolds number is small. The vortex shedding on the blunt body surface produces negative pressure behind the bluff body when the critical value is above the Reynolds number, and the pressure difference between the front and back becomes larger. At this time, the pressure drag becomes the main source of the resistance of the flow over the bluff body. For example, the pressure resistance of the large passenger and cargo aircraft in the cruise state accounts for 60–70% of the total resistance. The pressure difference resistance caused by the separation of the air flow at the rear of the car body even accounts for more than 85% of the total resistance while the passenger car body is running at a high speed [11] (Fig. 3.6).

The design on drag reduction has become an inevitable technical requirement for the problem of passing flow with the development of science and technology. As it is indicated in some data, the economic efficiency of fuel will be improved by 9% under mixed cycle conditions if the aerodynamic drag coefficient CD of Audi100 car is reduced from 0.42 to 0.30, and the economic efficiency of fuel can be improved by up to 25% when the car is driven at a constant speed of 150 km/h [12].

However, the shedding of vortex generated in the separation zone not only improves the pressure resistance of the bluff body, but also forms periodic pulsation forces on the surface of the structure. The pulsation force will cause periodic vibration of the bluff body when the bluff body is fixed by elastic support or allows elastic deformation, cause structural vibration, produce acoustic noise, and can also cause destructive vibration accidents, such as fatigue damage to buildings such as bridges, chimneys, tall towers, offshore platforms, and heat exchange stations. There will be a frequency lock phenomenon, resonance, or even galloping especially when the frequency of vortex shedding is close to the natural frequency of the bluff body structure, which raises the average resistance and significantly strengthens the pulsation force. The damage is more serious, and the typical flow-induced vibration brings structural damage [13–17].

Fig. 3.6 Flow resistance dominated by pressure drag. **a** Differential pressure resistance is the main air resistance that the vehicle is faced with; **b** Differential pressure resistance is the main source of resistance that the aircraft is faced with

It has been found in the practice that the control on the structure of the flow over the bluff body can not only reduce the resistance around the flow of the bluff body, but also suppress the structural vibration caused by the shedding vortex on the surface, which is of great significance to the reliability of the engineering structure. For example, the main beam of the Second Severn Crossing in the United Kingdom is a composition of steel and concrete. It was found that the structure of the bridge body had obvious vortex-induced vibration during the on-site inspection after the completion of the traffic. The engineer installed a spoiler under the main beam of the bridge [18] in order to adjust the coupling of the shedding of the vortex behind the bridge body and the resonance frequency of the main body of the bridge, which effectively avoided the vibration caused by the vortex shedding and improved the safety of the bridge.

Therefore, there is very important theoretical research significance and practical application value to explore and improve the flow structure on the surface of the bluff body. From the perspective of academic research, Roshko first measured the shedding period of the Carmen vortex at the tail of the bluff body since 1955 [1]. There are subsequent scholars who conducted a comprehensive study on the vortex shedding on the surface of the bluff body and the wake flow structure in the near field, many of which have also tried to change or inhibit the vortex shedding behind the bluff body with a series of control methods, and have achieved certain effects [19–26].

A systematic classification of flow control methods for bluff bodies are conducted based on current research, Choi et al. [27] For example, the control method is divided into passive control (with no energy input) and active control (with energy input) depending on whether energy is input into the flow field during the control process. In addition, different control methods can be divided into two-dimensional control (invariant along the span) and three-dimensional control (change along the span) according to the distribution characteristics of control disturbance source along the span (or azimuth).

Another example is that it can be divided into boundary layer control and wake control according to the control is to delay the separation of the boundary layer by acting on the boundary layer or directly in the wake flow. These control methods can be divided into open-loop control and closed-loop feedback control according to whether the control input is adjusted according to the control effect. Here, the classification of the bluff body flow control method is shown as the block diagram drawn in Fig. 3.7 according to the above classification criteria.

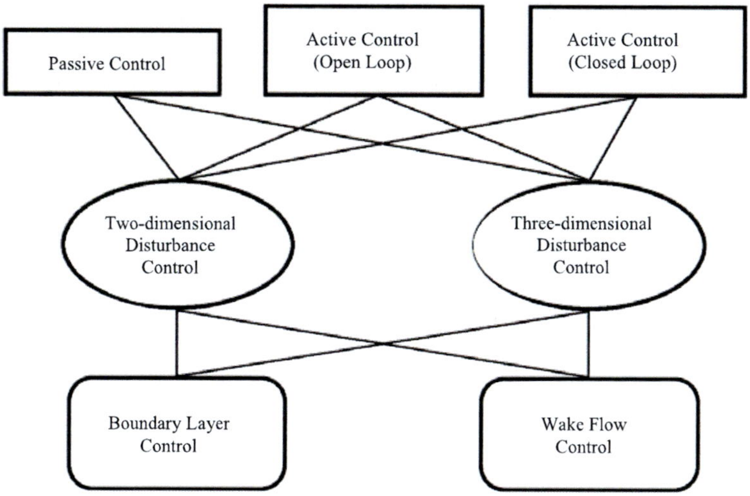

Fig. 3.7 Classification diagram of control methods for the flow over bluff bodies

3.2.1 Methods for Boundary Layer Control and Wake Flow Control

The flow control around the bluff body can be divided into boundary layer control and wake control according to the action area of the control target. Boundary Layer Control drives the transition of the boundary layer from laminar flow to turbulent flow by changing the flow characteristics of the boundary layer, thereby delaying the main separation and achieving resistance control. This control method is generally used for the flow control of the bluff body where the separation point changes with the Reynolds number, such as typical bluff bodies cylinders, spheres, etc. On the contrary, the wake flow control method is to directly act on the wake flow area to achieve the control of the resistance by changing the wake characteristics, so it can be applied in all bluff bodies with fixed and variable separation points [27].

3.2.1.1 Method for Boundary Layer Control

The mechanism of the boundary layer control method is the same as the that of "Drag Crisis" for the cylinder mentioned in Sect. 3.1 in the transition between the subcritical and supercritical area. The boundary layer changes from laminar to turbulent at this time, and it is surrounded by free flow. The fluid in high momentum enters the surface of the bluff body, which increases the flow momentum of the fluid near the wall area in front of the separation point, so that it can overcome the adverse pressure gradient at the rear of the bluff body, induce boundary layer separation and reattachment, and achieve delayed mainstream separation.

It has been shown in the tests that it is very effective for suppressing boundary layer separation to adopt boundary layer control for bluff bodies of which separation point varies with Reynolds number [28–31]. There are many methods to drive the boundary layer to enter the turbulent flow from the laminar flow state in advance, such as setting the vortex generator to increase the roughness of the object surface.

1. Roughness Control

It is an effective method to increase the surface roughness for the promotion of the laminar boundary layer to be converted into a turbulent boundary layer in advance. It is shown in Achenbach's rough ball test results that [28, 29], it can reduce the drag coefficient to properly raise the surface roughness within a certain range. Choi measured the velocity of flow on the surface of a rough sphere similar to a golf ball with tiny pits in detail in 2006.

It is shown in experimental results that the surface pits make the boundary layer on the surface of the sphere unstable, raise the turbulence intensity of the boundary layer, take up the momentum of the fluid near the wall, make the boundary layer reattach to the surface of the sphere after the first separation, and promote the delay of the main separation. The test data indicates that for rough balls with a pitted surface, its boundary layer enters from laminar to turbulent flow before the critical

Reynolds number. The drag coefficient C_D reaches the minimum value at the critical Reynolds number, then goes up rapidly as the Reynolds number increases, and the critical Reynolds number will decrease with the increase of roughness. In addition, the minimum drag coefficient of the sphere with larger roughness is also larger, where k is the depth of the pit or the height of roughness. Bearman carried out a wind tunnel test on a cylinder with pits, and a flow mechanism similar to a golf ball appeared, which can reduce the resistance of the flow around the cylinder [31].

2. Control on Micro-renovation

Micro-renovation on the surface of the bluff body is also a good method to control the boundary layer. For example, the tall tower 101 in Taipei was initially designed with a square cross-section. Concave corners were modified at the corner points to change the position of the flow separation point and reduce wind resistance and pulsating lift in order to reduce wind resistance and pulsating lift [32].

Hu et al. [33] trimmed and renovated the flake shapes for the corners of the square columns and carried out discussion on the effect on the flow. The photo shows the flow. It is shown in the results that the length of the cut edge directly affects the width of the trail, and compared with the square column, $C/d = 0.1$ (Projection Length of Cut Edge/Side Length of Square Column) the drag coefficient is reduced by 30%, which plays a better drag reduction effect. Wang et al. [34] carried out research on the effect of chamfering on the aerodynamic performance of rectangular columns (length/width = 2) in a pressurized wind tunnel. It is shown in the results that the Re-number is as high as 4×10^6, and the drag coefficient of rectangular columns with chamfered corners decreased from 0.75 before the modification to 0.35. In addition, some scholars carried out discussion on the influence of the rounded corner of the rectangular column on the aerodynamic performance of the flow. It is shown in the results that the average drag coefficient of the model decreases as the rounding rate increases, the average drag coefficient is more and more sensitive to the Reynolds number, and the characteristics of the flow field are affected by the rounds, which also drives the boundary layer to enter the turbulent flow earlier and delay the separation of the mainstream [35, 36].

3. Control on Groove/Rib

It is also an effective method for controlling the boundary layer to set a groove/rib structure on the surface of a bluff body, so as to achieve drag reduction [37, 38]. This method is inspired by the surface structure of shark skin and there had been studies shown that this non-smooth surface structure is beneficial to the reduction of the flow resistance of sharks.

Walsh [37] discovered that the flow to the surface of tiny ribs can effectively reduce the frictional resistance of the wall, who believed that the groove with drag reduction effect is a symmetrical V-shaped groove surface, on condition that the ribs must have sharp peak ridges there is a relationship between the spacing and low speed strip spacing, that is, the ribs have drag reduction performance when the dimensionless dimensions of the rib height h and the rib spacing s are $h \leq 25$ and $s \leq 20$, while the drag reduction effect is the best and the drag reduction rate can reach 8% when the rib size is $h = s = 15$. Launder and Li [38] carried out numerical calculations on the flow of smooth and non-smooth (L-shaped, U-shaped and V-shaped rib structures) channels, and also discovered that the V-shaped ribs had the best drag reduction, but the drag reduction effect for the rib size $h < 10$ is smaller.

The mechanism of surface groove/rib flow control is not clear at present, for which the more popular theory is the "Secondary Vortex Group" theory proposed by Bacher and Smith [39], that is, The secondary vortexes weaken the streamwise vortices and retain the low-speed fluid in the ribs when the opposite streamwise vortices interact with the secondary vortexes formed at the top of the rib, thereby inhibiting the formation and spreading of low-speed strips, which results in avoiding burst weakening and reducing frictional resistance.

4. 3D-Forcing

From the point of view of the effect on the above-mentioned boundary layer control methods, it mainly affects the boundary layer in two dimensions. There are many scholars also tried to set a varying control body along the span (azimuth) of the two-dimensional bluff body, in order to achieve three-dimensional control of the boundary layer. These structures that change along the span will promote the phase imbalance of the vortex in the wake, breaking the two-dimensional characteristic of vortex shedding, thereby increasing the base pressure C_{Pb} at the rear stagnation point of the bluff body, while attenuating the vortex strength of the bluff body. It is shown in the test results that this control method has a certain drag reduction effect.

For example, Zdravkovich set a spiral line on the span of the surface of the cylinder [40], and a serrated trailing edge [41] or a wavy trailing edge [42] on rear edge of round head blunt body. Bearman and Owen changed the straight axis of the cylinder along the span to a sinusoidal axis, and spirally connected the hemispherical protrusion to the surface of the cylinder [43], all of which achieved a good drag reduction effect.

The control effect on the flow around the structure is small for a square prism, a bluff body with a relatively fixed boundary layer separation point (it is generally considered that the boundary layer separation point is at the corner of the cylinder), if a 3D-Forcing disturbance source is added to the rear surface to control its flow structure. Therefore Bearman & Owen [43, 44] set the span ripples in front of the square column on the windward side, while the trailing edge still maintains the prism face and the undulations on the windward face can restrain the shedding of eddies.

Darekar [45] discovered that 3D-Forcing can cause deformation of the two-dimensional shear layer with numerical calculations at low Reynolds numbers, which makes it difficult for the upper and lower shear layers to attract and be involved in each other, forming Karman Vortex Street. It is also discovered in the research that, the width of the spreading arrangement is close to the spreading wavelength of mode A instability for 3D-Forcing, and its drag reduction effect is the best.

Park et al. [46] and Kim [47] installed small-scale fins near the separation point of the blunt body and the cylinder respectively, the combination of parameters will make the Karman vortex street in the near field complete by adjusting the height (l_y), width (l_z) and phase of the fins the span distance between adjacent ribs (λ),

The intensity has been significantly reduced even if it will be regenerated farther downstream, and a 33% drag reduction effect can be achieved. It is also further verified that the drag reduction effect is the best when the wavelength in 3D-Gradient arrangement is close to the spreading wavelength with mode A instability. In order to add the control effect of 3D-Forcing (The parameter combination is $l_y/b = 0.2$, $l_z/b = 0.2$, $\lambda/b = 0.2$, $Re = 4200$), Bearman and Brankovi [48] set up humps on the surface of the vibrating cylinder. It is found in the experiment that these humps attenuated their vortex shedding strength to a certain extent and suppressed the vortex-induced vibration intensity.

The boundary layer control methods listed above are all based on the idea of passive control. There is no artificial injection of energy into the flow field for these control methods, while another type of control method for boundary layer drag reduction on passing flow is based on the idea of active control, for which, energy is injected into the field to suppress vortex shedding and achieve a drag reduction effect.

It introduces several typical boundary layer active control methods as the following.

5. Active Rotation

It has been found in practice that the vortex shedding frequency will be correspondingly locked at a certain frequency when the bluff body is actively rotating at a fixed frequency. This phenomenon is called "Frequency Locking". The strength of the vortex shedding in the rear of the bluff body will be enhanced, the average resistance increases, and the pulse action is strengthened. There are some scholars for the other way based on this and successfully took this phenomenon to achieve drag reduction control. The typical example is that Tokumaru and Dimotakis [49] actively rotated the cylinder to achieve the suppression and resistance control of the falling vortex.

(Among them, $S_f = \frac{d}{U_\infty} f$ is the rotation frequency of the cylinder).

6. Active Vibration

Jeon et al. adopted high-frequency vibration on the sphere. It is found in the experiment that the vortex of the sphere will also "lock the frequency" in a certain vibration frequency domain. The separation of the boundary layer is delayed for the sphere under high-frequency vibration and the strength of the vortex has also been controlled. The boundary layer separation occurs at 80°–90° for smooth un-vibrated spheres, in which, b is the flow display of the sphere under vibration at a certain frequency, which indicates that the first boundary layer separation occurs at 105°–110°. And at 110°–115°, reattachment occurs, and finally the main flow shunt point occurs at 130°. As it is verified in the test results that the adoption of vibration in a specific frequency domain is conducive to the pressure recovery of the ball rear body and reduces the drag coefficient C_D.

7. Active Jet Flow

The active jet flow in the boundary layer is also an effective method for the active control of the boundary layer. The flow display under the control of the microjet is used before the boundary layer is separated. From the display of the flow, the micro jet flow effectively delays the flow separation of the mainstream. The recovery of pressure on rear body has also been improved.

Kim and Choi [50] adopted numerical calculation (Reynolds number range *Re* = 47–3900), and the suction and suction slits are arranged in the cylinder along the span. The slits have the characteristics of sinusoidal distribution along the span. The effect of the suction and suction is similar to that of the arrangement of 3D-Forcing disturbance. There is a phase mismatch between the vortex layers along the wingspan. The Karman Vortex Street almost disappears in the near wake flow area and the far field area will reproduce the weak Karman Vortex Street. It is shown in the calculation results that this sinusoidal suction control reduces cylindrical drag.

The adoption of blowing/suction in the boundary layer can also suppress the flow separation of the wing at a large angle of attack to a certain extent, for a streamlined body such as a wing. The airfoil of NACA 0012 adopted boundary layer suction control at an angle of attack of 18°, where L_j is the relative position of the suction position in the chord direction of the wing, θ is the angle between the jet and the wing surface, and A is the ratio of the jet velocity to the incoming velocity, which is the same as the suction velocity and under the angle, it is the separation control effects of different suction positions at the same suction speed and angle [51].

8. Zero Mass Jet Flow

Glezer et al. [52, 53] proposed a device with a suction function in the 1990s in addition to the above-mentioned boundary layer active jet control method. Vortex is generated at the edge of the orifice by the periodic movement of the flexible diaphragm in the sealed cavity. The device does not add fluid mass to the outflow, so it is also called zero mass jet flow, but it will input non-zero linear momentum into the outflow. A flow control effect of zero-mass jet is set in the cylindrical boundary layer. The zero-mass jet flow transforms the laminar boundary layer into a turbulent boundary layer. It is shown in the test results that it can effectively suppress the separation of the boundary layer and control the vortex shedding.

It was found through their experimental research that the drag reduction control effect with adoption of a zero-mass jet flow on the cylinder has a certain relationship with the installation position of the jet actuator, but it is not sensitive to the blowing frequency, and its optimal drag reduction effect can reach 25%.

3.2.1.2 Wake Flow Control Method

Several methods of drag reduction control by directly controlling the wake of the bluff body will be introduced in this section based on the control action zone, the control method of the wake layer is opposite to the control method of the boundary layer. These control methods can be divided into 2D control and 3D control based on the influence of 2d or 3D control on wake flow. It can be divided into passive control and active control depending on whether energy is injected into the flow field.

1. Passive Wake Flow Control

A typical two-dimensional control method on passive wake flow is to set a shunt plate or a secondary small cylinder on the rear surface or downstream of the bluff body. For example, Anderson and Szewczyk [54] installed a straight and wavy spreader plate behind a two-dimensional cylinder, and found that a spreader plate that satisfies a certain length can effectively suppress the formation of shedding vortex. The strength of the Karman Vortex Street in the wake flow is attenuated, which is beneficial to the cylinder. The pressure of the body recovers and its tail vortex formation length gets to be larger. Hwang et al. [55] calculated through simulation that, to set a splitter plate downstream of the cylinder can suppress vortex shedding and significantly reduce drag and the lift when $Re = 100$, and there is a best size and optimal installation position for the setting of the manifold.

Hamed et al. [56] added end plates to the upper and lower tail edges of the semi-cylinder. The end plates with different parameter combinations at the trailing edge have different effects on the flow field, and their drag reduction effects are also different. It is shown in the results that the drag reduction effect when $l/d_c = 1$ is the best, reaching 25%, and the drag reduction effect brought by $l/d_c = 0.25$ is only 3%.

It can also achieve a good control effect to Add a flexible divider on the rear surface or downstream of the cylinder. Kunze and Brucker [57] added a flexible flow field adaptive hairy small baffle on the rear edge of the cylinder. Hu et al. [58] added elastic trailing edge to the cylindrical trailing edge, which has achieved a certain effect of flow reduction and drag reduction control.

Strykowski and Sreenivasan [59] realized the direct control on wake flow by placing secondary small cylinders in the wake flow area of the cylinder. The Karman Vortex Street for reference cylinder can be well suppressed by the secondary small cylinders, and its resistance also got effective reduction. It is shown in the test results that the size of the effective area for the placement of the secondary small cylinder is affected by d/d_s (d is the diameter of the large cylinder and d_s is the diameter of the secondary small cylinder) and shrinks as d/d_s increases.

It is worth noting that the control efficiency of the above two-dimensional passive control method on wake flow is greatly affected by the incident angle of the incoming flow, and it can be found by comparing the research results of different literatures that, the control program of the partition plate used requires the size of the partition plate being larger, which limits the application of these control methods in actual engineering.

2. Active Wake Flow Control

Active wake flow control is to inject energy into the wake flow field in an active way to achieve control on wake flow. The basic idea is to apply momentum in the wake flow area to change the energy of the fluid in the local area and promote the overall wake structure to be controlled, so as to achieve resistance reduction of the passing flow.

Fransson et al. [60] set up a two-dimensional slit with suction function at the rear stagnation point of the cylinder. The cylindrical flow field structure significantly shortened the formation length of the tail vortex under the effect of suction for the structure of cylindrical flow field under different blowing and suction conditions. The result is opposite under the blowing condition. The proper suction function of the rear stagnation point has a certain drag reduction effect.

Qu et al. [61] set a suction slit at the rear stagnation point of a square prism, and acted on the wake area with a zero-mass jet (also called a synthetic jet). In addition, its drag reduction effect is significant when $f_e/f_0 = 6$ (where f_e is the actuation frequency and f_0 is the natural vortex off frequency of the square column). The experimental results also indicated that, the wake flow vortex shedding frequency will be consistent with the zero-mass jet frequency under certain parameter combinations, under the effect of the synthetic jet, and the wake will gradually change to a symmetrical state. It is shown in the results through comparison on the zero at the stagnation point behind the cylinder and the effect of mass jet control [62] that it is better for the flow over the bluff body with variable separation point.

Artana et al. [63] formed a high-pressure electrode on the rear surface of the cylinder with electromagnetic excitation, forming a plasma jet to achieve wake control. The wake flow of the cylinder and the electromagnetic excitation. The flow field structure has undergone significant changes under the influence of the plasma jet. It can be seen from the test results that electromagnetic excitation effectively promoted the recovery of the pressure coefficient on the rear surface of the cylinder.

Yu and Monkewitz [64] tried to heat the fluid in the cylindrical wake area to change the fluid density in the wake flow area. It is also shown in the results that the strength of the wake can be properly suppressed and the drag coefficient can be reduced by heating.

Each type of boundary layer and wake control that are introduced above, or whether they are active or passive, can be explained according to the stability theory proposed in the literature [65, 66]: The existence of Absolute Instability area is necessary for the vortex shedding on the surface of the bluff body [65, 66].

It is also indicated in many literatures that the development of global instability in wake flow is controlled by the local instability in the near-field area of wake flow. The boundary layer control and wake flow control are realized by eliminating or weakening the local absolute instability in the near-wake area to suppress the occurrence of global instability, thus suppressing the strength of Karman Vortex Street [67].

The mechanism has a certain relationship with the change of the local absolute instability if both of the small cylinders are placed in the wake to control the wake intensity. Strykowski and Sreenivasan [59] and Hwang et al. [55] set up two small cylinders to control the wake flow in the wake flow area. Strykowski got the effective control area through $d/d_s = 10$. The locally unstable area of the cylinder in the near-field area is on the shear layer and wake flow center line separated above and below the cylinder.

Fransson et al. [60] reduced the fluid density near the wake area with heating to suppress the increase of local instability, and Leu and Ho [67] reduced the length of the absolutely unstable region near the wake through blowing/inhaling at the tail of the plate, thus suppressing the growth of global instability. The active boundary layer control on the jet flow similar to 3D-Forcing adopted by Kim and Choi [50] can also be interpreted through the instability theory, and the disturbance of its specific wavelength reduced the growth rate of local instability. And the optimal 3D-Forcing control wavelength predicted through the theoretical analysis of stability is about 3d, which is consistent with the experimental conclusions from Kim and Choi. It is noteworthy that, the theory of instability is based on the function of the average velocity profile, on condition that, it must provide enough energy to change the velocity profile of the wake flow.

The premise to control the absolute instability effectively is to provide sufficient energy to change the wake velocity profile. A blow/suction speed equivalent to the free incoming flow speed is required to suppress vortex shedding if the drag/drop method at the rear stagnation point is used to achieve drag reduction.

3.2.2 Active Closed-Loop Control

Each kind of control method mentioned above is to achieve control on the flow field and resistance over the bluff body through multiple experiments to find a combination of control parameters with positive effects. In general, the control input is relatively fixed, and the control system belongs to the open-loop working mode without self-adjusting according to the flow field. Active closed-loop control methods with flow field feedback signals have been put into practice and application with the development of micro-electromechanical systems and control theory and the improvement of computer performance. Such closed-loop control methods can stimulate control according to the real-time control effect of the flow field and make real-time adjustments to improve and increase control efficiency. The advantage of the active closed-loop control method over the open-loop control method is that it obtains the response information of the flow system to the input excitation, so that it can obtain better control performance than the open-loop control method, and the control efficiency is relatively high, which is the hot spot in volume flow control research [68].

Closed-loop control can be divided into the following categories according to different feedback signals.

3.2.2.1 Active Control with Single Linear Feedback

It is generally believed for the wake control of a cylinder that the effective control performance is that, the wake flow of the cylinder is still symmetrical even in the subcritical area, and the characteristics of the Karman Vortex Street are not prominent. The shear layer generated by the separation of the boundary layer on the surface will first destabilize and attract each other when the Reynolds number of the flow over the cylinder reaches the critical Reynolds number, gradually forming the Karman Vortex Street. It should be able to suppress this flow instability under the critical Reynolds number for an effective closed-loop control method.

For specific applications, Berger [69] first made use of the signal of the hot wire sensor (velocity) located in the wake flow to control the movement of the bimorph on the surface of the cylinder, who opened the idea to control the flow structure around the cylinder with a closed-loop control method. Williams and Zhao [70] carried out feedback control on the speaker actuation based on the velocity phase information at the wake when $Re = 400$. It is shown in the results that the vortex shedding frequency and velocity fluctuations on the cylinder surface are reduced.

Roussopoulos [71] also carried out a similar feedback control test at a low Reynolds number. The results show that the hotline for collecting feedback signals is placed in the near wake area to suppress vortex shedding when $Re = 65$, but it will fail because the feedback signal is too weak if the hotline is moved to the far field Zone. The sensor can still clearly detect the eddy current signal at about cylinder diameter of 9 in downstream. The vortex shedding cannot be achieved well if it is used as the feedback signal excitation input. The possible reason for this control is that the feedback signal has a serious lag. It is worth noting that this control method does not perform well under high Reynolds numbers.

Park et al. [72] simulated the wake control of a cylinder with a suction slit by using a numerical method in terms of numerical calculation, and controlled the suction action of the slit with the signal of the sensor located in the wake area as a feedback control signal.

The blowing-suction in the slit follows $\phi_{upper}\ (t) \sim v(x_s,\ y = 0,\ t)$ and $\phi_{lower}\ (t) \sim -v\ (x_s, y = 0, t)$, in which ϕ_{upper} and ϕ_{lower} indicate The blowing/suction speed in the upper and lower slits respectively, and $v(x_s, y = 0)$ is the lateral speed of the center line at the sensing position x_s. The resistance coefficient changes with the sensor position, which indicates that the feedback sensor position x_s has a great effect on the control effect. The vortex shedding at $x_s = 2.75$ is completely suppressed. It also shown in the calculation results that the control method has a poor control effect at a high Reynolds number [72].

The speed signal in the shear layer can also be taken as a feedback input to achieve a good control effect in addition to taking the speed signal in the wake area as a feedback control signal. Huang measured the speed at the upper shear layer with hot wire technology, and took this as a feedback control signal to drive the loudspeaker located on the upper surface of the cylinder with a reverse phase offset method to achieve better control performance [73].

The overall shedding vortex strength can be suppressed although this scheme is controlled only on one side of the cylinder, which also shows that the vortex shedding is the result of the instability of the two parallel shear flows above and below the cylinder.

Zhang et al. [74] adopted a proportional integral derivative (PID) control method for the flow control of a bluff body in a vibration state to suppress the flow around a square column under resonance supported by a spring.

A piezoelectric ceramic actuator is installed on the upper and lower surfaces of the square column. There are three different feedback signals adopted to control the piezoelectric ceramic actuator, respectively, the flow velocity signal (PID-u control), the flow structure oscillation signal (PID-Y control) and the combination of the two signals (PID-Yu control). It is indicated in the results that the feedback control based on PID-Yu shows the best performance, almost completely suppresses the generation of eddy currents, and the square column under the resonance condition returns to static.

3.2.2.2 Control Based on Reduced Order Model (ROM)

Combined with the above description, a single linear feedback based on a single sensor can achieve good performance at low Reynolds numbers, because the bluff body wake is often caused by an unstable mode leading to the formation of Karman Vortex Street at low Reynolds numbers, while when it is at high Reynolds numbers, there are many unstable factors of wake, and multiple feedback signals are needed to accurately judge the flow field.

A feasible method is to reduce the model reduction (ROM) and the basic idea of reduced order reduction is based on appropriate orthogonal decomposition (POD), point vortex model or other order reduction methods to reduce the dimensionality of the flow field information and provide a low-order feedback signal, in order to reduce the number of feedback sensors, but to ensure that important dynamic flow field signals can be obtained, so as to achieve a simpler and more feasible control method than that based on the full flow field signal [12], which also opened up ideas for its engineering applications.

Its basic idea is to decompose high-energy modes (corresponding to large-scale coherent structures) with POD and low-energy modes from the flow field [75], and artificially take the signal that is cut off after the low-energy mode as the flow field feedback control input signal. For example, in practice, Gillies [76] obtained a simplified wake model (ROM) with the POD method, and made use of a neural control network to determine the control input, who successfully achieved the suppression of vortex shedding on the cylindrical surface at a higher Reynolds number.

Graham et al. [77] obtained the ROM based on the POD method when $Re = 100$, and applied it to the control of the flow around the cylinder, who achieved a good control performance, but the control accuracy during the control process will decrease with the development of the regulation time. Therefore, it is necessary to reset the ROM in stages to raise the control accuracy, which may be related to the cutting off of information of the low-order mode many times, which causes the flow field model to be too simple, making it invalid. Bergmann et al. [78] proved with a similar method that the feedback control through POD-based ROM is similar to the feedback control with the complete Navier–Stokes equation, and further proved that the POD-based ROM is effective as a feedback input. Sexuality.

The ROM can also be obtained with the point vortex model when the flow field is basically controlled by the vortex in the wake [27]. Cortelezzi [79] performed nonlinear feedback control on the flow around a vertical plate and a cylinder, and set a suction point on the downstream wall of the plate or cylinder, which can adjust the blowing and suction action according to the flow rate conditions, form and separate a pair of similar vortex in the boundary layer. It is shown in the result that it inhibits the formation of wake flow vortex.

Henning et al. [80] also set up an actuating device on the rear surface of the bluff body with a low-dimensional vortex model (ROM) to generate small-structure vortex to control the flow on the model surface and obtain a good drag reduction performance.

3.3 Optimal Control and Suboptimal Control

Optimal control theory has caught more attention in the field of flow control. Hintermuller and Hinze [81] applied optimal control theory to analyze many fluid mechanics problems, and the optimal control conditions are given for various physical situations, however, the optimal control algorithm they proposed also has limitations, because the speed information of the flow field is required in the derivation process to solve the conjugate equation, from which to obtain the required feedback control signal. However, it is not practical for the three-dimensional Navier–Stokes equation.

He et al. [82], Protas and Styczek [83], Flinois and Colonius [84] and Li and Aubry [85] controlled the rotation of the cylinder with numerical simulation based on optimal control theory to realize wake flow control. Among them, He et al. [82] proposed an optimal feedback control method based on the optimal control theory model. It is shown in the numerical calculation results that the cylinder is at a certain amplitude within the range of Reynolds number $Re = 200$–1000. The active flow direction vibration can achieve the effective control of the wake flow under the condition of the vibration frequency and achieve a drag reduction benefit of 30–60%.

Poncet et al. [86] applied a 3D-Forcing disturbance along the span of the cylinder based on the optimal control theory, which generates a positive suppression effect. The structure of cylindrical wake flow was between Mode A and Mode B, and drag reduction control was achieved.

What cannot be overlooked is that the above control attempts all make use of the minimum cost function based on the velocity of the laminar boundary layer. This algorithm requires information on a full flow field of the global time period and its optimization time interval is greater than the vortex shedding interval, which also determines that it is currently impossible to directly apply the optimal control method to the actual situation, and can only achieve eddy current control at a low Reynolds number [27].

Choi et al. [87] introduced a suboptimal feedback control algorithm, which seeks optimal conditions in a short period of time to avoid the long iteration process required for global optimal control, in order to overcome the limitations of the optimal control theory proposed by Berger [69].

This algorithm on control has been successfully applied to cylinders [88], back steps [89, 90] and spheres [91] wake flow control.

Min and Choi [88] controlled the flow over the cylinder with application of suboptimal feedback control at $Re = 100$ and 160. They installed a feedback sensor that controls the suction boundary layer on the cylindrical surface. The control input is optimized to make the cylindrical vortex shedding weaken and disappear based on the difference between the real-time pressure on the cylindrical surface and the preset pressure, and the pulsation intensity of average resistance and lift resistance is significantly reduced.

Jeon and Choi [91] made use of numerical simulation for the flow around the sphere to achieve a certain control effect in $Re = 425$ with the sub-optimal control method. The difference between the real-time pressure coefficient on the surface of the sphere and the pressure coefficient distribution under ideal conditions is controlled by blowing and sucking on the surface of the sphere. The structure wake flow is well suppressed.

The key factor of optimal and suboptimal control is to determine the cost function to be minimized (such as drag or lift), and its control effect also depends on the selection of cost functions. For example, the drag reduction effect when the cost function is defined as the difference between the actual and ideal surface pressure coefficients is better than that when the drag coefficient difference function is used as the cost function [88], because the difference function on surface pressure coefficients between the actual and ideal is a non-linear function, which has stronger adaptability.

3.4 Summary and Outlook

It mainly introduces different flow control methods in the first two subsections, and classifies and illustrated the different control methods. People have achieved certain results in flow control in the long-term practice, but there are still some problems at present to be solved.

1. Limitations of Reynolds Number

The flow structure on the surface of the bluff body will show different states with the changes of the Reynolds number. Therefore, it is also a concern whether the effective control method within a specific Reynolds number range is effective in other Reynolds number ranges. From the perspective of the current research, the universal control method is still to be developed.

For example, the linear feedback control from Park et al. shows that, when it is at a low Reynolds number $Re = 6$, it can completely suppress the vortex shedding, but it is not effective at higher Reynolds numbers [72]. The essential reason is that the surrounding flow field at higher Reynolds numbers contains global instability different from low Reynolds numbers, so a simple feedback control method based on single-sensor measurement cannot achieve wake control at high Reynolds numbers. Another example, changing the surface roughness of the bluff body in the sub-critical Reynolds number ($10^4 < Re <$ critical Reynolds number) range can achieve reduction resistance control [28, 29], but it is ineffective at lower Reynolds numbers since the boundary layer flow within this Reynolds number range is absolutely stable. On contrary, control methods that change local/global instability seem to be applicable to different Reynolds numbers, such as back-stagnation point jet flow, three-dimensional geometric modification [46], and 3D-Forcing control [50].

There is another thing worth noting that we usually define the Reynolds number based on the free flow velocity and the characteristic height h of the bluff body, while another important characteristic scale for separation flow is the thickness θ of the boundary layer before separation. For example, Petrusma and Gai [41] discovered that the eddy current dynamics in the wake flow are also affected by the flow characteristics of the boundary layer before separation, such as the back pressure coefficient C_{Pb} of a two-dimensional bluff body with a blunt trailing edge decreasing with h/θ rapidly increasing, (but for $h/\theta > 40$, C_{Pb} is almost constant, which is −0.6 to −0.5) [37]. Therefore, the thickness θ before the boundary layer separation may also be an important parameter that needs to be considered in the future development of flow control technology.

2. The Problem for Minimum Drag Coefficient

What is the minimum achievable drag coefficient for control on drag reduction of bluff body? This is a question worth discussing since the minimum drag coefficient will limit the performance of drag reduction control, while which also provides direction for the development of effective control methods and is an important research topic in the future.

The vortex shedding during the flow around the bluff body will significantly increase the average resistance and the pulsating strength of the lift compared with no vortex shedding. Therefore, it is generally assumed that that the resistance when the vortex shedding does not occur in the tail of the bluff body by control means is the minimum drag coefficient. However, this flow still has a large shape resistance in fact, so it is very efficient if a flow control method can be developed that can achieve the distribution characteristic of the surface pressure coefficient of the surface of the bluff body similar to that of the potential flow (with zero shape resistance). For example, Min and Choi's goal of control is to reduce the difference between the true and ideal pressure coefficient distribution on the cylindrical surface [88].

3. Efficiency of Control

The drag reduction efficiency is an important factor in the case of active control, and it is highly dependent on the control method adopted. The control method with high efficiency should be developed as small energy input as possible in the future control method research to realize the maximum control effect, which is also an important problem that needs to be solved urgently in the future when the active control method is put into practical engineering applications [92–98].

From the perspective of the current development level of drag reduction control, it is still in constant exploration, however, the future efficient drag reduction control method must be based on the integration of multiple disciplines with the development of micro-electromechanics, new material technology, etc.

References

1. Roshko A (2012) On the wake and drag of bluff bodies. Springer Berlin Heidelberg
2. Bearman PW (1967) On vortex street wakes. J Fluid Mech 28(4):625–641
3. Griffin OM (1981) Universal similarity in the wakes of stationary and vibrating bluff structures. J Fluids Eng 103(1):607–617
4. Williamson CHK, Brown GL (1998) Aseries in 1/Vre to represent the Strouhal-Reynolds number relationship of the cylinder wake. J Fluids Struct 12(8):1073–1085
5. Roshko A (1993) Perspectives on bluff body aerodynamics. J Wind Eng Ind Aerodyn 49(1):79–100
6. Williamson CHK, Roshko A (1990) Measurements of base pressure in the wake of a cylinder at low Reynolds numbers. Zeitschrift Fur Flugwissenschaften Und Weltraumforschung 14(1):38–46
7. Tropea C, Yarin L, Foss JF (2007) Springer handbook of experimental fluid mechanics. Springer, Berlin Heidelberg
8. Schewe G (1983) On the force fluctuations acting on a circular cylinder in crossflow from subcritical up to trans-critical Reynolds numbers. J Fluid Mech 133(133):265–285
9. Zurong D (2003) Fluid mechanics (Middle Volume). Higher Education Press, Beijing
10. Roshko A (1961) Experiments on the flow past a circular cylinder at very high Reynolds number. J Fluid Mech 10(3):345–356
11. Peng G (2015) Research on technology of aerodynamic drag reduction for vehicles based on tail flow structure. Jilin University, Changchun
12. Yu Z (2006) Automobile theory, 4th edn. Mechanical Industry Press, Beijing
13. Li H, Laima S, Ou J et al (2011) Investigation of vortex-induced vibration of a suspension bridge with two separated steel box girders based on field measurements. Eng Struct 33(6):1894–1907
14. Yang B, Gao F, Wu Y etc (2006) Experimental research on vortex-induced vibrations of submarine pipeline near seabed boundary in ocean currents. China Ocean Eng (English Edition) 20(1):113–121
15. Mukundan H, Hover FS, Triantafyllou MS (2010) A systematic approach to riser VIV response reconstruction. J Fluids Struct 26(26):722–746
16. Nguyen T, Koide M, Yamada S et al (2012) Influence of mass and damping ratios on VIVs of a cylinder with a downstream counterpart in cruciform arrangement. J Fluids Struct 28(1):40–55
17. Trim AD, Braaten H, Lie H et al (2005) Experimental investigation of vortex-induced vibration of long marine risers. J Fluids Struct 21(3):335–361
18. Macdonald JHG, Irwin PA, Fletcher MS (2002) Vortex-induced vibrations of the Second Severn Crossing cable-stayed bridge—full-scale and wind tunnel measurements. Struct Build 152(2):123–134
19. Williamson CHK (1996) Vortex dynamics in the cylinder wake. Annu Rev Fluid Mech 28(1):477–539
20. Coutanceau M, Defaye JR (1991) Circular cylinder wake configurations: a flow visualization survey. Appl Mech Rev 44(6):255
21. Berger E, Wille R (2003) Periodic flow phenomena. Annu Rev Fluid Mech 4(1):313–340
22. Lin JT, Pao YH (2003) Wakes in stratified fluids. Annu Rev Fluid Mech 11(1):317–338
23. Bearman PW (2012) Vortex shedding from oscillating bluff bodies. Annu Rev Fluid Mech 16(1):195–222

24. Oertel J (2003) Wakes behind blunt bodies. Annu Rev Fluid Mech 22(1):539–562
25. Griffin OM (1991) Review-vortex shedding lock-on and flow control in bluff body wakes. J Fluid Eng 113(4)
26. Williamson CHK, Govardhan R (2004) Vortex-induced vibrations. Annu Rev Fluid Mech 36(1):413–455
27. Choi H, Jeon WP, Kim J (2008) Control of flow over a bluff body. Annu Rev Fluid Mech 40(1):113–139
28. Achenbach E (1974) The effects of surface roughness and tunnel blockage on the flow past spheres. J Fluid
29. Achenbach E (1971) Influence of surface roughness on the cross-flow around a circular cylinder. J Fluid Mech 46(2):321–335
30. Jin C, Jeon WP, Choi H (2006) Mechanism of drag reduction by dimples on a sphere. Phys Fluids 18(4):112
31. Bearman PW, Harvey JK (2012) Control of circular cylinder flow by the use of dimples. Aiaa J 31(10):1753–1756
32. Irwin PA (2008) Bluff body aerodynamics in wind engineering. J Wind Eng Ind Aerodyn 96(6):701–712
33. Hu JC, Zhou Y, Dalton C (2006) Effects of the corner radius on the near wake of a square prism. Exp Fluids 40(1):106–118
34. Wang X, Gu M, Quan Y (2016) Research on the Reynolds number effect of the aerodynamic coefficient of a two-dimensional rectangular cylinder for the width-to-thickness ratio of the fillet treatment of 2: 1. Eng Mech 33(1):64–71
35. Maxworthy T (1969) Experiments on the flow around a sphere at high Reynolds numbers. J Appl Mech 91(3)
36. Dean B, Bhushan B (2010) Shark—skin surfaces for fluid—drag reduction in turbulent flow: a review. Philoso Trans R Soc Math Phys Eng Sci 368:4775–4806
37. Walsh MJ (2012) Riblets as a viscous drag reduction technique 21(4):485–486
38. Launder BE, Li SP (1993) On the prediction of riblet performance with engineering turbulence models. Flow, Turbul Combust 50(3):283–298
39. Bacher E, Smith C (1985) A combined visualization-anemometry study of the turbulent drag reducing mechanisms of triangular micro-groove surface modifications. Springer Berlin Heidelberg
40. Zdravkovich MM (1981) Review and classification of various aerodynamic and hydrodynamic means for suppressing vortex shedding. J Wind Eng Ind Aerodyn 7(2):145–189
41. Petrusma MS, Gai SL (1994) Effect of geometry on the base pressure recovery of segmented blunt trailing edges. Aeronaut J 98(977):267–274
42. Bearman PW (1997) A study of three-dimensional aspects of vortex shedding from a bluff body with a mild geometric disturbance. J Fluid Mech 330(1):85–112
43. Bearman PW, Owen JC (1998) Reduction of bluff-body drag and suppression of vortex shedding by the introduction of wavy separation lines. J Fluids Struct 12(1):123–130
44. Owen JC, Bearman PW, Szewczyk AA (2001) Passive control of viv with drag reduction. J Fluids Struct 15(3–4):597–605
45. Darekar RM (2001) Flow past a square-section cylinder with a wavy stagnation face. J Fluid Mech 426(426):263–295
46. Park H, Lee D, Jeon WP et al (2006) Drag reduction in flow over a two-dimensional bluff body with a blunt trailing edge using a new passive device. J Fluid Mech 563(563):389–414
47. Kim D (2014) Control of circular cylinder wake using three-dimensional momentum forcing. Oil Prod South Inst Acad Life 5:13–15
48. Bearman P, Brankovi CM (2004) Experimental studies of passive control of vortex-induced vibration. Eur J Mech/B Fluids 23(1):9–15
49. Tokumaru, Dimotakis (1991) Rotary oscillation control of a cylinder wake. J Fluid Mech 224(224):77–90
50. Kim J, Choi H (2005) Distributed forcing of flow over a circular cylinder. Phys Fluids 17(3):97

51. Lebeau RP, Huang L, Huang PG et al (2004) Numerical study of blowing and suction control mechanism on naca0012 airfoil. J Aircraft 41(5):1005–1013
52. Smith BL, Glezer A (1998) The formation and evolution of synthetic jets. Phys Fluids 10(9):2281–2297
53. Glezer A, Amitay M (2002) Synthetic jets. Annu Rev Fluid Mech 34(34):503–529
54. Anderson EA, Szewczyk AA (1997) Effects of a splitter plate on the near wake of a circular cylinder in 2 and 3-dimensional flow configurations. Experi Fluids 23(2):161–174
55. Hwang JY, Yang KS, Sun SH (2003) Reduction of flow-induced forces on a circular cylinder using a detached splitter plate. Phys Fluids 15(8):2433–2436
56. Hamed AM, Vega J, Liu B et al (2017) Flow around a semicircular cylinder with passive flow control mechanisms. Exp Fluids 58(3):22
57. Kunze S, Brucker C (2012) Control of vortex shedding on a circular cylinder using self-adaptive hairy-flaps. CR Mec 340(1–2):41–56
58. Hu Y, Pan C, Wang JJ (2014) Vortex structure for flow over a heaving cylinder with a flexible tail. Experi Fluids 55(2):1682
59. Strykowski PJ, Sreenivasan KR (1990) On the formation and suppression of vortex 'shedding' at low Reynolds numbers. J Fluid Mech 218(218):71–107
60. Fransson JHM, Konieczny P, Alfredsson PH (2004) Flow around a porous cylinder subject to continuous suction or blowing. J Fluids Struct 19(8):1031–1048
61. Qu Y, Wang J, Sun M et al (2017) Wake vortex evolution of square cylinder with a slot synthetic jet positioned at the rear surface 812:940–965
62. Feng LH, Wang JJ (2010) Circular cylinder vortex-synchronization control with a synthetic jet positioned at the rear stagnation point. J Fluid Mech 662(7):232–259
63. Artana G, Sosa R, Moreau E et al (2003) Control of the near-wake flow around a circular cylinder with- electrohydrodynamic actuators. Exp Fluids 35(6):580–588
64. Yu MH, Monkewitz PA (1990) The effect of nonuniform density on the absolute instability of two-dimensional inertial jets and wakes. Phys Fluids A 2(7):1175–1181
65. Huerre P, Monkewitz PA (1990) Local and global instabilities in spatially developing flows. Ann Rev Fluid Mech 22(1):473–537
66. Chomaz JM (2004) Global instabilities in spatially developing flows: non-normality and nonlinearity. Annu Rev Fluid Mech 2(37):357–392
67. Leu TS, Ho CM (2000) Control of global instability in a non-parallel near wake. J Fluid Mech 404(404):345–378
68. Gad-el-Hak M (2000) Flowcontrol: passive, active, and reactive flow management. Cambridge University Press, Cambridge
69. Berger E (1967) Suppression of vortex shedding and turbulence behind oscillating cylinders. Phys Fluids 10(9):S191–S193
70. Williams JEF, Zhao BC (1989) The active control of vortex shedding. J Fluids Struct 3(2):115–122
71. Roussopoulos K (1993) Feedback control of vortex shedding at low Reynolds numbers. J Fluid Mech 248(248):267–296
72. Park DS, Ladd DM, Hendricks EW (1994) Feedback control of von Ka'rma'n vortex shedding behind a circular cylinder at low Reynolds numbers. Phys Fluids 6(6):2390–2405
73. Huang XY (1996) Feedback control of vortex shedding from a circular cylinder. Exp Fluids 20(3):218–224
74. Zhang MM, Cheng L, Zhou Y (2004) Closed-loop-controlled vortex shedding and vibration of a flexibly supported square cylinder under different schemes. Phys Fluids 16(5):1439–1448
75. Balachandar S (1998) Turbulence, coherent structures, dynamical systems and symmetry. Cambridge University Press, Cambridge
76. Gillies EA (1998) Low-dimensional control of the circular cylinder wake. J Fluid Mech 371(371):157–178
77. Graham WR, Peraire J, Tang KY (1998) Optimal control of vortex shedding using low-order models. Part I-open-loop model development. Int J Numer Methods Eng 44(7):945–972

78. Bergmann M, Cordier L, Brancher JP (2005) Optimal rotary control of the cylinder wake using proper orthogonal decomposition reduced-order model. Phys Fluids 17(9):93–178
79. Cortelezzi L (1996) Nonlinear feedback control of the wake past a plate with a suction point on the downstream wall. J Fluid Mech 327(327):303–324
80. Henning L, Pastoor M, King R et al (2007) Feedback control applied to the bluff body wake. Springer, Berlin Heidelberg
81. Hintermuller M, Hinze M (2002) On some control problems in fluid mechanics. Esaim Math Model Numer Anal 36(4):725–746
82. He JW, Glowinski R, Metcalfe R et al (2000) Active control and drag optimization for flow past a circular cylinder. J Comput Phys 163(1):83–117
83. Protas B, Styczek A (2002) Optimal rotary control of the cylinder wake in the laminar regime. Phys Fluids 14(7):2073–2087
84. Flinois TLB, Colonius T (2015) Optimal control of circular cylinder wakes using long control horizons. J Fluid Mech
85. Li F, Aubry N (2003) Feedback control of a flow past a cylinder via transverse motion. Phys Fluids 15(15):2163–2176
86. Poncet P, Hildebrand R, Cottet G et al (2008) Spatially distributed control for optimal drag reduction of the flow past a circular cylinder. J Fluid Mech 599:111–120
87. Choi H, Temam R, Moin P et al (1993) Feedback control for unsteady flow and its application to the stochastic Burgers equation. J Fluid Mech 253(-1):509
88. Min C, Choi H (1999) Suboptimal feedback control of vortex shedding at low Reynolds numbers. Phys Fluids 401:123–156
89. Jeon S, Choi J, Jeon W et al (1999) Active control of flow over a sphere for drag reduction at a subcritical Reynolds number. J Fluid Mech 517:113–129
90. Kang S, Choi H (2002) Suboptimal feedback control of turbulent flow over a backward—facing step. J Fluid Mech 463
91. Jeon S, Choi H (2005) Suboptimal feedback control for drag reduction in flow over a sphere. Springer Berlin Heidelberg
92. Barkley D, Henderson RD (1996) Three-dimensional Floquet stability analysis of the wake of a circular cylinder. J Fluid Mech 322(322):215–241
93. Lin JC, Towfighi J, Rockwell D (1995) Instantaneous structure of the near-wake of a circular cylinder: on the effect of Reynolds number. J Fluids Struct 9(4):409–418
94. Norberg C (2003) Fluctuating lift on a circular cylinder: review and new measurements. J Fluids Struct 17(1):57–96
95. Cantwell B, Coles D (1983) An experimental study of entrainment and transport in the turbulent near wake of a circular cylinder. J Fluid Mech 136(136):47
96. Yamagishi Y, Kimura S, Oki M (2010) Flow characteristics around a square cylinder with changing chamfer dimensions. J Visual 13(1):61–68
97. Thompson MC, Leweke T, Williamson C (2001) The physical mechanism of transition in bluff body wakes. J Fluids Struct 15(3):607–616
98. Jeon S, Choi J, Jeon WP et al (2004) Active control of flow over a sphere for drag reduction at a subcritical Reynolds number. J Fluid Mech 517(517):113–129

Chapter 4
A Functional Based Prediction Method for Hypersonic Aerodynamic Force and Heat

4.1 Introduction

The aerospace industry has experienced more than 100 years of development since the birth of the aircraft in 1903, and mankind have created milestones one and another in the history of flight. Basic theories, testing methods, and calculation methods on flight have been gradually developed and improved with the development of the times and industry, which have always played an irreplaceable role in opening up one new era after another for incompressible subsonic velocity, compressible subsonic velocity, transonic velocity, and supersonic velocity. Qian Xuesen started to carry out basic research as early as the 1930s on hypersonic flow and established a similarity law for hypersonic flow under certain conditions. Von Karman and others had realized in the 1940s that the era of hypersonic flow in the future was coming. There has been a series of research advances over the past 80 years such as rockets, missiles, space shuttles, return capsules, wave riders and gliders which have shown humanity the dawn of the era of hypersonic flight. However, from the perspective of the development of science, the development for aerodynamics theory that is compatible with the era of hypersonic flight is far less than the accumulation of theories corresponding to other eras, which shows to a certain extent that humans are still far away from the era of hypersonic flight.

Hypersonic aircrafts are an important means for precision strikes and strategic deterrence in the future within ultra-long range and even global range, which are closely related to the national economy and national security, and are the focus of aerospace research in countries around the world. The predictive capability on aerodynamic force/heat marks the development level of hypersonic aircrafts in a country. It can provide important guidance for the structural layout, functional module layout, and flight control design of aircrafts to predict the aerodynamic/thermal performance indexes of the aircraft in flight with ground test data. Designers often need to leave a greater margin for flight control and heat protection in the case of insufficient

M. Zhao, *Prediction and Validation Technologies of Aerodynamic Force and Heat for Hypersonic Vehicle Design*, Springer Aerospace Technology,
https://doi.org/10.1007/978-981-33-6526-1_4

predictive capabilities, which will cause the aircraft to be cumbersome and affect performance.

Therefore, predictions on aerodynamic force/heat under hypersonic flight conditions are still faced with huge challenges, which are urgent scientific research problems to be solved. In the nature of physics, the difficulty for prediction is that the quantitative change of flight speed brings a qualitative change in the content of aerodynamic research: from subsonic to supersonic, and then to hypersonic, qualitative change occurs for the flow medium on the outer edge of the aircraft with the quantitative change of flight speed, the gas molecules have undergone vibrational excitation, dissociation or even ionization, and ordinary air becomes a complex medium that continuously undergoes thermo chemical reactions. The microscopic chemical and physical changes of this medium deliver an important effect on the macroscopic motion state through changing the thermodynamics and transport characteristics of the medium, which shows characteristics of non-linear, non-equilibrium and multi-scale flow, while the traditional flow similarity ratio may no longer be applicable, which has brought great challenges to the accurate prediction on aerodynamic force/heat. The United States used 44 wind tunnels once during the development of the space shuttle with the wind tunnel test time more than 90,000 h, but the partial prediction error of aerodynamic heat still exceeded 100%. The requirements on shape and structure, especially for advanced new types of aerospace aircrafts, are becoming more and more precise, and the aerodynamic layout and structural design are becoming more and more difficult. Therefore, the difficulty in predicting characteristics of the aerodynamic force/heat of hypersonic flow has become an important bottleneck with constraint on the research and development of aerospace aircrafts.

Hypersonic flow is coupled with gas physics phenomena at the same time such as shock wave dynamics, thermodynamic and thermo chemical reactions. The complete physical model description is still lacking so far on the effect of turbulence, transition, shock wave/boundary layer interference, high temperature real gas effect, etc.

The results from theoretical analysis and CFD calculation deviate greatly from the prediction results on aerodynamic force/heat of complex aircraft. Flight tests are also faced with difficulties such as high cost, long cycle time, difficult measurement, and lack of data. The wind tunnel test is still the most reliable data acquisition method for prediction on aircraft aerodynamics at present, and it is also the main basis for verifying other research methods (Fig. 4.1).

For this purpose, the United States, Europe, China and other major aerospace countries have established their own equipment for hypersonic ground tests. There are conventional hypersonic wind tunnels that obtain high Mach numbers by reducing static temperature and sound speed, and hypersonic wind tunnels that directly heat the air, as well as a high-enthalpy shock wave wind tunnel with strong shock compression. These wind tunnels have made great progress with different methods to improve the Mach number, total temperature and flow velocity of the test airflow, thus providing a lot of data for hypersonic flow and aircraft design. However, it is difficult to completely reproduce the flow field state under actual hypersonic flight conditions due to the limitations of various wind tunnel driving capabilities and heating methods and the different degrees of pollution of the test airflow (that is, the

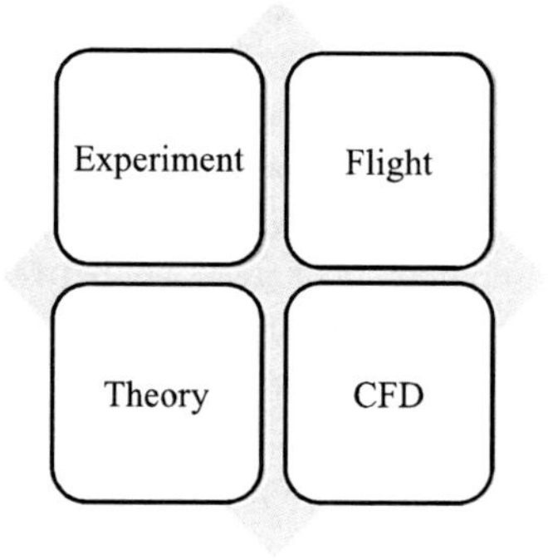

Fig. 4.1 Four means of research on hypersonic aerodynamic force/heat

flight Mach number Ma, Reynolds number Re, total temperature T°, total pressure P., total temperature/wall temperature ratio T_0/T_W, flow field pattern, density ratio in front and rear of shock wave, flow field pattern, geometric size of test model, relative height of bulge, boundary layer and other characteristic parameters), and the flight condition parameters can only be partially simulated. It leads to a significant difference between the ground test and flight conditions, and even the difference between the simulated physical mechanism. The direct availability is poor of the test results. The results obtained from ground wind tunnel tests cannot generally be directly applied to aircraft design and flight test planning. It can be seen from the point of view of mathematics that, the use of wind tunnel tests to predict the flight status is essentially an extrapolation due to the lack of ground test capabilities. Therefore, traditional modeling methods that are suitable for interpolation are not applicable such as data fitting/interpolation/neural network/support vector machine.

At the same time, there are differences in the flow conditions that different types of equipment for hypersonic ground test can simulate. The test conditions between different wind tunnels under the same Mach number also have significant differences in key parameters such as Reynolds number, static temperature, total pressure, and flow speed, which makes the obtained wind tunnel test data short of comparability, which brings great uncertainty on the prediction of aerodynamic performance and the study of the gas flow law of high-enthalpy thermal chemical reaction.

It is of great significance for the development of hypersonic flight technology and understanding the laws of high-temperature gas flow to conduct research on related theories and related methods of hypersonic aerodynamic force/heat through effective use of wind tunnel test data, explore the influence law of the difference between the wind tunnel test state and real flight conditions, and establish a predictive method from wind tunnel test characteristics data to flight conditions, so as to improve the reliability and applicability of wind tunnel test data.

4.2 Overview of Prediction Method

The prediction on hypersonic aerodynamic force/heat is to mine and analyze the wind tunnel test data under certain ideas and theoretical guidance, establish the connection of test data between different wind tunnels, develop correlation methods based on important parameters, and form universal physical laws and associated parameters that can describe complex flow phenomena. Then, prediction method on hypersonic aerodynamic force/heat is obtained based on these physical laws and related parameters to get more reliable data on aerodynamic characteristics of the aircraft under specific flight conditions, so as to support the research and development of new aircrafts.

Aerodynamic force is a whole quantity, and its value is the reflection of the integral effect of the forces of various parts of the aircraft in different measurement directions. Accordingly, as a local quantity, aerodynamic heat depends on the local flow environment and is proportional to the temperature gradient, so aerodynamic heat is a differential effect. From the perspective of mathematics, the solution of aerodynamic heat value is much more difficult, and the difficulty of solution differs by two orders of magnitude. In fact, the local state of flow is closely related to the surface curvature and characteristics of the aircraft. It is generally believed that there is no geometric similarity criterion for the aerodynamic heat characteristics of the aircraft. The aerodynamic heat prediction of full-scale aircraft under flight conditions cannot be predicted from the wind tunnel test data of the shrinkage model, so the research on aerodynamic heat prediction is more complicated.

At present, the methods for hypersonic aerodynamic force/heat prediction can be divided into the following types.

1) Geometric Simplification Method

This method first simplifies the complex aircraft into a few simple standard models, then carries out comprehensive analysis with aerodynamic force/heat equations of simple geometry, and makes corrections through wind tunnel tests, so as to extrapolate to flight conditions for prediction.

This method is simple and effective, but it ignores the interaction between the geometric flow fields. Obviously, effective aerodynamic heat prediction is not possible in the action area, and aerodynamic force prediction may also bring great errors.

2) Theoretical Simplification Method

This method obtains the so-called key factors under certain assumptions through physical analysis, which ignores the "minor" factors, establishes an engineering experience equation that reflects the aerodynamic/thermal law, and predicts aerodynamic performance. For example, it is supposed based on the Stanton number St that, $\frac{St Re_x^n}{(\rho^*/\rho_e)^{1-2n}} = C$, in which Rex is the local Reynolds number; p* is the density corresponding to the reference enthalpy; Pe is the gas density at the outer edge of the boundary layer, and C is a constant. It is assumed based on the correlation between the ground and flight data that, $\frac{\frac{q}{(q_{ref})_{FLT}}}{\frac{q}{(q_{ref})_{WT}}} = C$, in which q is the wall heat flux density (the heat flux density in the following chapter is referred to as heat flux for short); q_{ref} is the reference heat flux, $(\cdot)_{FLT}$ represents the quantity in flight state, $(\cdot)_{WT}$ represents the amount under the wind tunnel state.

Based on physical analysis, this method is easy to understand. It is of great significance for the rapid estimation of aerodynamic force/heat. However, on the one hand, it will affect the prediction accuracy since the equation ignores the secondary factors under certain assumptions. On the other hand, the selection of primary and secondary factors has certain empirical and special cases. The equation can be misleading when the ignored factors gradually strengthen or even when dominant.

3) Comparing Calculation Method

This method corrects the calculation results according to a certain proportional coefficient according to the difference between CFD simulation, wind tunnel test and flight test results, so as to predict the thermal environment under real flight conditions [1, 2].

This method is easy for operation, but it is essentially equivalent to packing all factors without consideration of CFD and CFD errors (including grids, turbulence models and other factors) into a black box and adjusting it with a coefficient, so it lacks the physical and mathematical basis and the reliability of the results has yet to be verified.

4) Key Difference Correction Method

This method is adopted to carry out semi-empirical correction and extrapolation for key differences (single factor). For example, Elsenaar [3], Haines [4, 5] and others carried out research on the influence of the Reynolds number of wind tunnel tests due to the difference in aircraft model shrinkage ratio and flow parameters. Poisson-Quinton [6], Elsenaar et al. [3] carried out research on the effects of real gas effects.

Based on physical data results, this method has certain reliability, and is easy to understand. However, there is only one factor considered at a time, and it is not easy to generalize to multi-factor situations. It may need to examine the interaction of multiple factors simultaneously for the actual aerodynamic force/heat.

5) Mathematical Approximate Model Method

Regarding aerodynamic force/heat prediction as a mathematical problem, this method is carried out based on aerodynamic force/heat data, with hypersonic aerodynamic force/heat coefficient as the objective function, through pure mathematical methods for direct approximate modeling such as interpolation, fitting/regression, artificial neural network, support vector machine, etc., and prediction with the approximate model.

This type of methods can be divided into two stages for the development: It needs to assume that the objective function satisfies a certain function structure and is subjective for early approximation models. For example, Peterson [7] carried out research on the wind tunnel/flight aerodynamic data correlation with interpolation (extrapolation) of the supersonic cruise aircraft XB-70-1.

Rufolo et al. [1] proposed for association of test aircraft FTB-1 with a polynomial model after logarithmic transformation (with form like $g(Re) = a{\cdot}\log(Re)^2 + b{\cdot}\log(Re) + c$). Rufolo et al. adopted the one-dimensional power function model and carried out analysis on the aerodynamic data of the test aircraft PRORAUSV 1 by nonlinear fitting. Morelli et al. [8] carried out analysis on X43-A test data with stepwise regression and orthogonal function modeling methods. De Priest et al. [9] performed rapid aerodynamic thermal analysis of the NASA reusable launcher RLV with five statistical modeling methods, such as Kriging based on the heat flow of the ball head stagnation point.

The approximation models on later stage have less requirements on the objective function structure and are more objective. For example, Wang Qing et al. [10] identified missile aerodynamic parameters and optimize input design with maximum likelihood estimation. Lee et al. [11] also analyzed the estimation accuracy of flight derivatives with the maximum likelihood estimation method. Norgaard et al. [12], Rajkumar et al. [13], Malmathanraj and Tyson [14] adopted neural network prediction methods. Ravikiran and Ubaidulla [15] predicted the aerodynamic drag coefficient with support vector machines.

This type of methods solved the problem according to the idea of "Physical Data *T* Mathematical Algorithm-Physics Prediction", with a weakly correlated "Sandwich Layer" between physics and physics. Generally, more ideal prediction (interpolation) results can be obtained when the point to be predicted is the inner point of the envelope of the known data point. On contrary, the prediction deviation increases rapidly when away from the inner point, which will result in the performance of prediction (extrapolation) is unacceptable.

6) Method of Forecast Based on Functional [16]

Based on functional intelligent optimization algorithm, this method is the combination and development of theoretical simplification method and mathematical approximate model method. It achieves the simplification of intelligent theory by using the function of automatic parameter selection and combination of functional optimization. It optimizes the function structure and coefficients at the same time with the use of functional optimization can achieve the ability, so that mathematical approximate modeling is performed to obtain physical laws and associated parameters, thus reflecting a certain universality for hypersonic aerodynamic force/heat prediction. This method provides a new research idea for aerodynamic force/heat prediction: physical data-mathematical algorithm + physical knowledge T-physical laws T-physical prediction.

Hypersonic aerodynamic data analysis based on functional has become a new trend of aerodynamic thermal prediction methods with the development of artificial intelligence, which has broad application prospects and, therefore, is the recommended method in this chapter. It will introduce the idea of this method, algorithm implementation, preliminary results, and discuss its development prospects in the following sections.

4.3 An Intelligent Data Correlation Method Based on Functional Optimization

4.3.1 Historical Inspiration

It requires to start with the analytical and invariant laws of aerodynamics/heat so as to make fundamental progress in hypersonic aerodynamic force/heat prediction. It can be inspired through the discovery history of Kepler's laws. Both "Geocentric Theory" and "Heliocentric Theory" believed that the orbit of a celestial body was an absolute circle (the most perfect and harmonious circular motion) in the sixteenth century.

However, there are always deviations in the results of the prediction based on the circular motion (not large, easy to be ignored, for example, the semi-major axis of the earth and lunar orbit is 384,400 kms; the short semi-major axis is 379,700 kms, the difference between the two is 4700 kms, the error is about 1.2%). Tycho (1546–1601), a Danish astronomer, has observed 750 stars in one day for thirty years, who recorded changes in their relative positions, corrected errors in the catalog, and improved data accuracy. Kepler found that the celestial body did not do circular motion, but elliptical motion through 8 years of hard work with the basis of the observation data of Tycho. In addition, on this basis, he found that the area swept by the unit time is equal and there is a constant relationship $R^3/T^2 = K$ between the revolution period T of the planet and the semi-major axis R. Kepler's laws turned a new page in astronomical history.

To give up accuracy is to give up the law. Perhaps there would be no celestial mechanics today if Tycho, Kepler and others were satisfied with the error of 1.2% at that time and did not do in-depth research. Nowadays, the amount of data accumulated on hypersonic aerodynamic force/heat is constantly increasing, and the accuracy of the data is also constantly improving. It has come to a critical period from quantitative change to qualitative change to obtain regular conclusions. There will be no fundamental progress in aerodynamic force/heat research if we are only satisfied with the results and accuracy of engineering algorithms [17].

An excavation of law requires the support of artificial intelligence. As of today, the analytical expression of Kepler's laws on planetary motion is actually not complicated. However, Kepler obtained the law through 8 years of manual deductions and calculations (some said 4 or 12 years) at that time. It is not that Kepler is not smart enough, but it is human weakness to find that the underlying law behind the "disorganized" data. As the saying goes, "Great law is most simplified, while it is easy to understand but difficult to discover." To make matters worse, the aerodynamic force/heat law of hypersonic aircrafts may be much more complicated than Kepler's laws on planetary motion, which is difficult to obtain by manual deduction. Fortunately, the rapid development of artificial intelligence over the recent years has made it possible to replace manual deduction with functional intelligent optimization.

4.3.2 Basic Idea

4.3.2.1 Multi-space Correlation Theory [16]

Various types of hypersonic wind tunnels have appeared with the development of hypersonic ground test equipment. The simulation capabilities are still very different even with the same kind of wind tunnel, which increases the diversity and disorder of the test data [18]. It is not enough to meet the needs of advanced aircraft design for the analysis still according to the traditional theoretical simplified analysis method with only one or two key parameters (such as Mach number/Reynolds number) selected manually. However, the test result is still physical if viewed from a higher dimension, although the wind tunnel test state is not on the flight plane. Therefore, intrinsically related are the aerodynamic force/heat characteristics obtained by the wind tunnel test and the corresponding characteristics under the flight state (Fig. 4.2). It contributes to the prediction of the flight status as long as the result of each test is reliable, while the contribution is different, which depends on the test capability of the wind tunnel and the level of test design. Free flow velocity, static temperature, static pressure, model reduction ratio and wind tunnel aerodynamic parameters constitute a complete multi-dimensional full-parameter test data space for a certain aircraft shape and flight attitude. There is a restriction relationship H (T, P) among the static temperature,

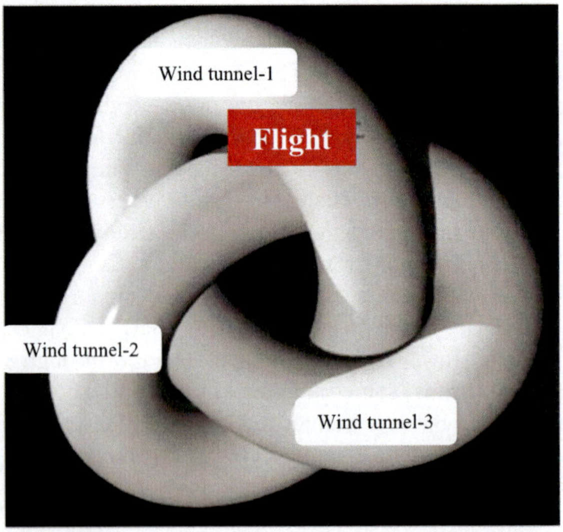

Fig. 4.2 Schematic diagram of the internal correlation of test data under different conditions

static pressure and altitude of the flight corridor, which constitutes the flight subspace in the test data space.

The whole space has very high dimensions, but there are only a few dimensions with a strong direct correlation with the objective function (i.e., aerodynamic force/heat coefficient). At the same time, there are some dimensions also with a strong correlation, so the related relationship between these dimensions and objective function can be indirectly reflected through the related relationship with other dimensions. Some of the dimensions have an interactive relationship that affects other dimensions together, or affects the objective function together. It is extremely difficult to analyze and express the laws of physics directly in the high-dimensional full space. Therefore, dimensionality reduction must be done in an appropriate way to analyze the aerodynamic/thermal parameter dependence. That is to say, the algorithm should deal with many multidimensional subspaces after dimensionality reduction, as multi-space for short. The prediction method with intelligent optimization based on functional is to calculate and evaluate separately in multiple spaces, so as to obtain the correlation between different parameters/parameter transformations and aerodynamic force/heat.

4.3.2.2 Functional Space Analysis Method

One must stand at a global level to perform hypersonic aerodynamic force/heat predictions according to multi-space related theory, so as to break through traditional thinking, and discover physical laws in a wider space. The first is to carry out analysis on the problem in a multidimensional space that can be reduced in dimension (Fig. 4.3a): The traditional analysis method is to analyze in a space with a certain number of dimensions preset. The new method points out that it needs to start from a multi-dimensional full parameter space and adopt appropriate dimensionality reduction method for analysis. On the other hand, it is necessary to deal with the problem

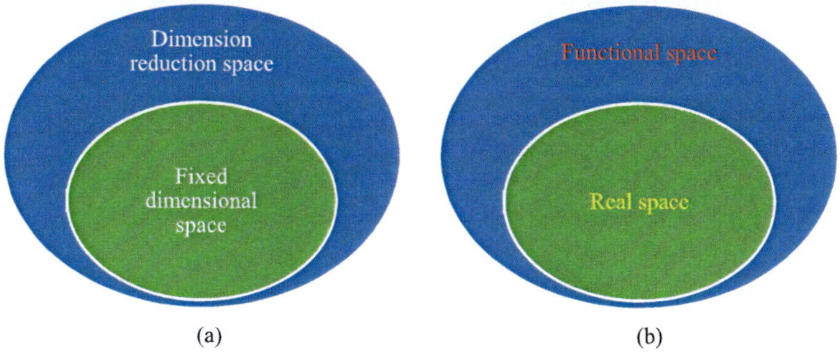

Fig. 4.3 The spatial inclusion relationship between the functional spatial analysis method and the traditional analysis method **a** dimensional relationship; **b** functional relationship

Table 4.1 Comparison of the functional space analysis methods and the existing methods

	Existing methods	Functional space analysis method
Spatial dimension	Single parameter, multi-parameter, fixed dimension	Full parameter space, adaptive dimensionality reduction
Modeling method	Assumption, empirical modeling	Functional space optimization
Modeling process	Manual approximation and deduction	Automatic optimization
Prediction methods	Extrapolation through change trends	Extrapolation through invariant laws
Prerequisites	All parameter ranges have been effectively simulated and the test results are reliable	All physical mechanisms have been effectively simulated and the test results are reliable
Scope of application	Envelope of test data points	Envelope of physical mechanism for experimental simulation

of the consistency of the world standing on the functional space rather than the real number space (Fig. 4.3b):

The traditional processing methods is to make empirical assumptions about the structure of the correlation function, such as obeying the power function relationship, exponential function relationship, polynomial relationship, hierarchical linear relationship (neural network method), etc., and then optimize the coefficients in the real space, while the new method points out that it is necessary to optimize the selection of parameters, function structure and undetermined coefficient set in the functional space at the same time, and realize the complexity control on the correlation function (Table 4.1).

To be more specific, taking aerodynamic heat as an example, we obtained various types of wind tunnel data on the ground. The types of wind tunnels include conventional hypersonic wind tunnels, heated air wind tunnels, high enthalpy shock tunnels, etc. The goal is to perform a consistency analysis of the sky and the earth with the data of these wind tunnels (hereinafter referred to as wind tunnel groups), and predict the aerodynamic heat environment of the aircraft when flying in the space. In fact, different types of wind tunnels can simulate different flow parameters and reflect different physical mechanisms. Therefore, every wind tunnel has its own data distribution law, which can be regarded as a partial expression of the overall law (Fig. 4.4a). The physical law on the consistency of aerodynamic force/heat between ground and space conditions as long as we can find out the uniform invariant laws (or laws understood as laws) that the heat environment of aircrafts abides by in different types of wind tunnels if we regard the flight test as a special ideal wind tunnel. In other words, the analysis on the consistency of the aerodynamic heat between ground and space tests is to explore the uniform invariant laws that the heat environment of the aircraft abides by in different types of wind tunnels based on the analysis of the change trend of heat flow under the combined influence of multiple parameters (Fig. 4.4b).

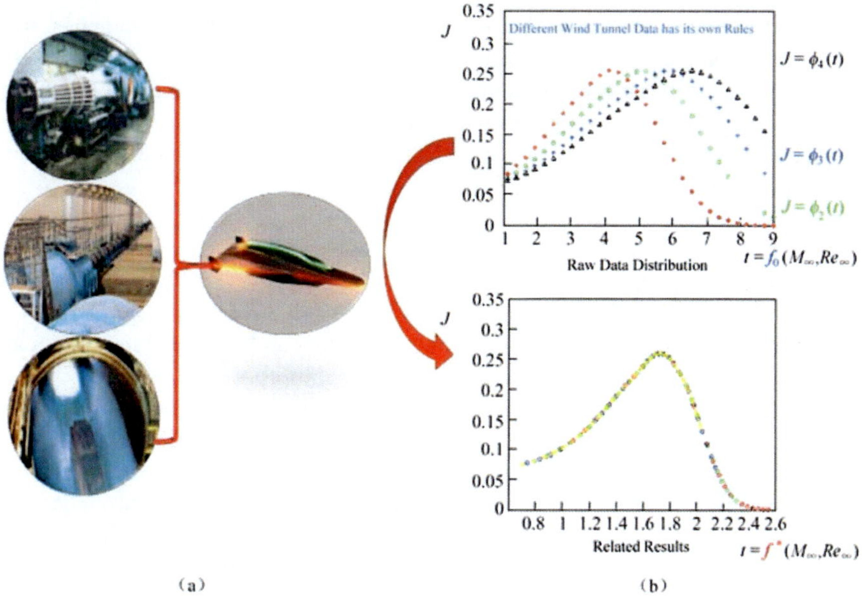

Fig. 4.4 Functional space analysis method for prediction of aerodynamic force/heat **a** the relationship between different wind tunnels and flight conditions; **b** The relationship between individual laws and invariant laws

It is noted that a function that represents the laws of physics is a point in the functional space. Therefore, the theoretical equation obtained by the simplified analysis of traditional physics can be regarded as a special case of the new method.

4.3.2.3 The Greatest Truths are the Simplest

The greatest truths are the simplest, which is an important thought of Chinese Taoist philosophy. It points out that the greatest principles (referring to basic principles, methods and laws) should be simple. There are similar thoughts in the West, which believe that Newton's greatest contribution to social thought in modern times is to "Solve the Mystery of Nature with Simple and Beautiful Mathematical Equation," and point out that any correct theory is simple in form and very useful with versatility. This coincides with the thought that The Greatest Truths are the Simplest from China.

The thought that The Greatest Truths are the Simplest is to require that the complexity of the model must not be too large for a model that describes hypersonic aerodynamic force/heat law (whether it is a trend for changes or an invariant law). That is to say, the true physical laws are generally not too complicated and relationships that are too complicated (such as the fitting relationship that implies many coefficients, including spline fitting, multivariate polynomial fitting, neural network, support vector machine, etc.) are often not a physical law, but just an appearance, a product of blindly catering to data.

To say the least, it is difficult to obtain them based on the current level of human knowledge if there are some extremely complex and highly nonlinear physical laws. In other words, the simplicity of physical laws themselves is an inherent factor in the feasibility of functional space analysis methods.

4.3.2.4 Change and Invariability

The understanding of human on the laws of nature can be divided into two realms: The cognition of the laws of change and the cognition of the laws of invariability. Mastering laws of invariability is the final goal to understand the world. The purpose to research on phenomena is to grasp the invariant laws of phenomena. In a certain sense, to grasp the invariant law is equivalent to grasp the essence of things. It is an effective research method to grasp the essence of the research object through invariant laws. For example, an ellipse is said to be a flattened circle. This is a phenomenon. The distance from the point on the ellipse to the center point is decreasing and increasing when moving towards the periphery. This is the trend of change. Only find the "The Sum of the Distances between the Two Fixed Points is Constant" or "The Ratio of the Distance to the Fixed Point and the Distance to the Fixed Line is Constant e, and $0 < e < 1$", can it be to grasp the essence of the ellipse, which raises the understanding of the ellipse to a higher level. We can predict the rest of the ellipse from a part of the ellipse with any one of these two invariant relations (Fig. 4.5).

The discovery of the physical laws of hypersonic through test data can also be divided into two levels (Fig. 4.6): The first level is to obtain the aerodynamic force/heat change trend through the quantitative relationship of the data, and the other higher level is to obtain the aerodynamic force/heat invariance law contained in it through analysis.

There is often certain stability in physically invariant laws (which can be characterized by certain similar parameters). The invariant law still holds when no new

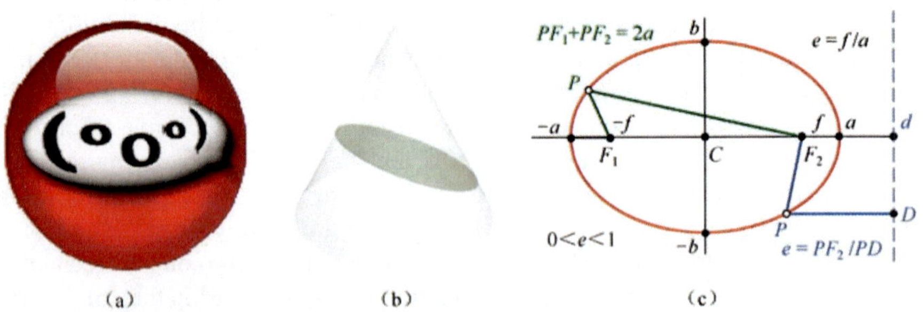

Fig. 4.5 The appearance and invariance relationship of the ellipse **a** ellipses of various appearance; **b** generation of ellipses; **c** invariant relations implied by ellipses

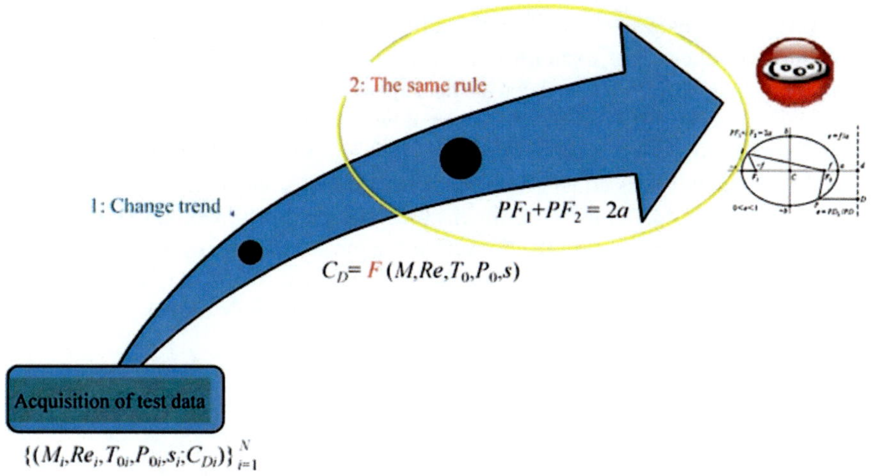

Fig. 4.6 Two levels of physical law discovery

physical phenomena (such as air vibration excitation, dissociation, etc.) appear and dominate, so there is higher reliability for the data association and prediction through the invariant law.

At the same time, there may be multiple invariant laws in hypersonic flow as in the previous elliptical example, but we can solve the problem of using the ground test data to predict the aerodynamic force/heat performance of the aircraft in flight as long as we can grasp one of these invariant laws.

In the following, how to use the test data of different wind tunnels to obtain the hypersonic aerodynamic force/heat law is explained from the two aspects of analyzing the changing law and the invariant law, so as to predict the flight status.

4.3.3 Functional Prediction Method for Analyzing Change Laws

4.3.3.1 Basic Principles to Realize Algorithm

First, let's go over the dimensional analysis. Mathematics can be said to be a black and white world without dimensions if you look at physics as a color world with dimensions.

Dimensional space establishes a bridge between physics and mathematics. The influence of dimensionality is generally considered in the derivation of physical equations (the process of law exploration), that is to say, a dimensional analysis is carried out. Dimensional analysis can determine the compatibility of physical equations in dimensional space, and thus check the completeness of equations. Of course, this is not absolute, and some useful equations do not have such compatibility and completeness. For example, the well-known plate heat flow equation: $q_w = \rho_\infty^N v_\infty{}^M C$, in which, q_w is the wall heat flow; P_∞ is the density of the incoming flow; v_∞ is the velocity of the incoming flow; the power function exponents M and N meet the laminar flow $M = 3.2$, $N = 0.5$, and the turbulent flow $M = 3.37$, $N = 0.8$. Obviously, this equation does not have dimensional compatibility. However, we believe that equations with dimensional compatibility should better reflect the physical laws of the system itself under the same complexity conditions.

The most famous work in dimensional analysis should be Buckingham's π theorem, which chooses the appropriate m basic parameters to make all parameters dimensionless:

$$f(x_1, x_2, \ldots x_m, x_{m+1}, x_{m+2} \ldots x_n) = 0 \tag{4.1}$$

so as to obtain a dimensionless relation that satisfies dimensional compatibility:

$$F(\pi_1, \pi_2 \ldots \pi_{n-m}) = 0 \tag{4.2}$$

π theorem solves the following problems:

(1) The selection criteria of basic physical quantities were that the selection of basic physical quantities must comply with the irrelevance in the dimensional space, $|A| \neq 0$;
(2) It can be dimensionless, so as to make sure that the equation does not have compatibility problems in dimensional space;
(3) It can reduce the number of independent variables (dimensionality reduction).

But π theorem leaves the following problems at the same time:

The selection of basic parameters in the full parameter space. In fact, the selection of basic parameters is not unique. Different selection methods can obtain different results, in which some may obtain useful results while some results may not make much sense.

The determination of the functional relationship after non-dimension. In fact, non-dimension can only guarantee the compatibility of dimensions, while how to obtain the appropriate functional relationship to represent the physical laws of the system is actually still an optimization process. Appropriate functional relationships can succinctly reflect the internal laws, while function relationships in too low complexity ignore too many factors and affect accuracy and function relationships in too high complexity may contain non-physical pseudo-laws.

It can be seen that the n-theorem provides a good idea to obtain the laws of physics, and it is the basic principle to realize the algorithm for the analysis laws and change laws. However, the theorem failed to completely solve the analytical expression of physical laws. Master insight is required in parameter selection and equation deduction to obtain a useful equation through dimensional analysis.

4.3.3.2 Functional Intelligent Optimization Algorithm

We propose an artificial intelligence algorithm—analytic matrix optimization algorithm (PME), in order to realize the adaptive selection of parameters and the automatic deduction of equations, and to solve the unresolved problem of n theorem. The algorithm can realize the adaptive dimensionality reduction of the full parameter space in the functional space, and detect the analytical expression of the function reflecting the flow law through use of hypersonic aerodynamic test data, with aerodynamic force / heat coefficient as the target.

The algorithm for analytic matrix optimization adopts an analytic matrix coding method, which overcomes the constraints of the tree structure (Fig. 4.7) of general genetic programming [19] (Fig. 4.7) on the programming language, and also solves The problem of excessive compression of information of the linear structure of Grammatical Evolution [20] (Fig. 4.8), which has been greatly improved in terms of the efficiency and stability of algorithm.

The basic idea of the PME algorithm is that the analytic expression of function reflecting physical law is decomposed into bases according to parameters, operators, coefficients and so on, then form an analytical relationship segment according to the genetic method, and finally form different genes into chromosomes, which is corresponding to an independent function expression and participating in the evaluation of candidate individuals in evolutionary algorithms. A typical example is listed as follows:

Coding: Map Function Space to Matrix Space. The function space is the target space, and the matrix space is the computing space of the search algorithm, which is an expression form easily recognized by the computer.

Base: The basic unit of analytical relationship, such as + , −, *x*, + , pow, sgn, sinh, exp, x_1, x_2, x_3, λ_1, λ_2, etc.

Gene: Analyze the Relationship Fragments, such as $(\lambda_1 + \sinh(x_i)\ \mathrm{sgn}\ (x_2))^2$.

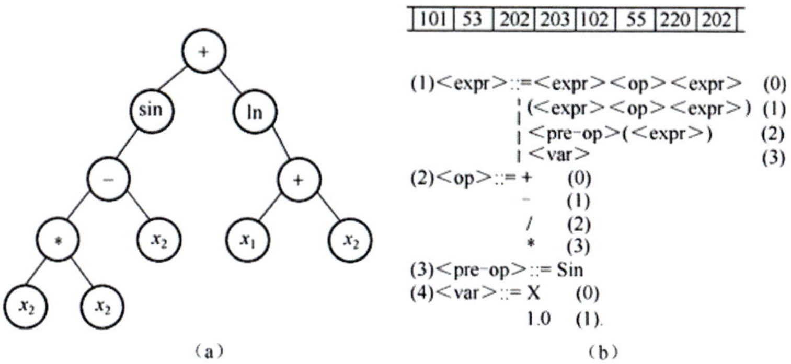

Fig. 4.7 Coding structure of existing algorithms, **a** genetic programming tree coding; **b** grammatical evolution linear coding

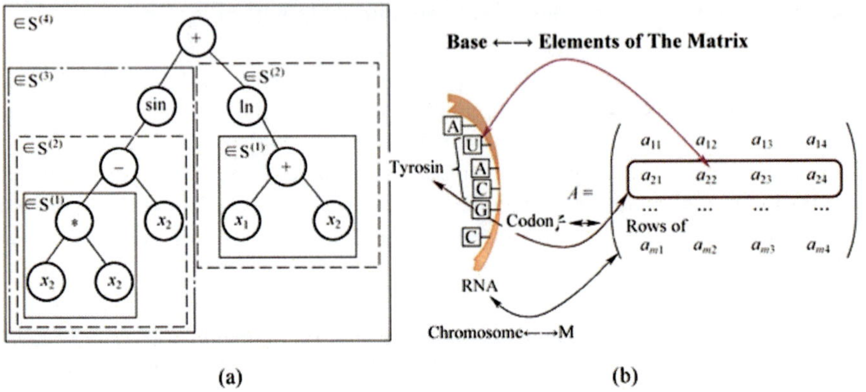

Fig. 4.8 Layered structure and bionic meaning of PME algorithm coding **a** PME algorithm coding layered structure; **b** bionic meaning (right)

Chromosome: Candidate Analytical Relationship Equation for Evaluation, such as $(\lambda_1 + \sinh(x_1)\,\mathrm{sgn}(x_2))^2/x_3$.

On this basis, the evolutionary algorithm is applied to perform the optimization calculation in the matrix space according to the following steps (Fig. 4.9), so as to obtain the optimal global analytical expression:

Initialization: some functions/matrices are randomly generated or some functions/matrices are manually input.

Evaluation: individual fitness is given in the evaluation space with the test data results.

Evolution: Intersection, variation, and the creation of new individuals.

Seeking the Best: "Select the Superior and Eliminate the Inferior".

Among them, the evolution process to generate new individuals can be completed by the evolution operator in Fig. 4.10. An efficient global optimization algorithm LDSE [21] is embedded in the evolution process to quickly optimize the coefficients in the analytical function, in order to speed up the evolution process.

The meanings of related symbols are as follows in the process of detecting the hypersonic aerodynamic force/heat conversion law:

Fi: Candidate function is an associated model of data consistency in different wind tunnels;

Mi: the corresponding integer matrix of the candidate function in the coding space, $Mi \in Z^{m\times n}$.

fit *(Mi):* fitness function, measurement candidate function characterizes the level of data consistency of different wind tunnels.

The intelligent optimization algorithm of the functional space can obtain an analytical relationship with global optimization characteristics in the multidimensional space with dimensionality reduction.

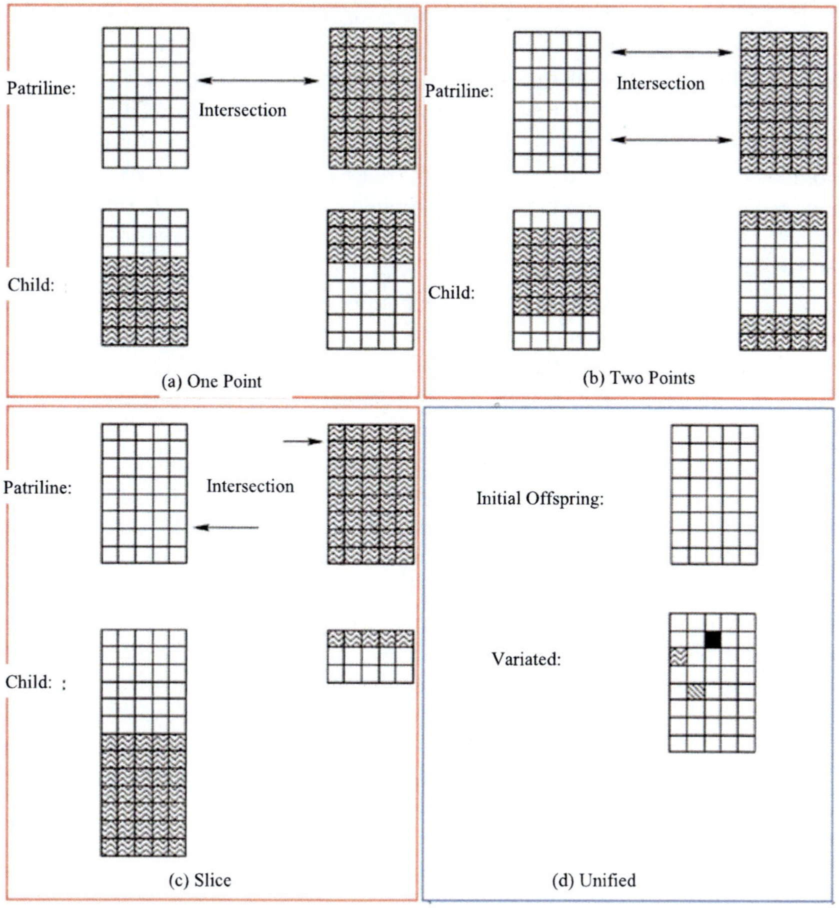

Fig. 4.9 Evolutionary operator of PME algorithm

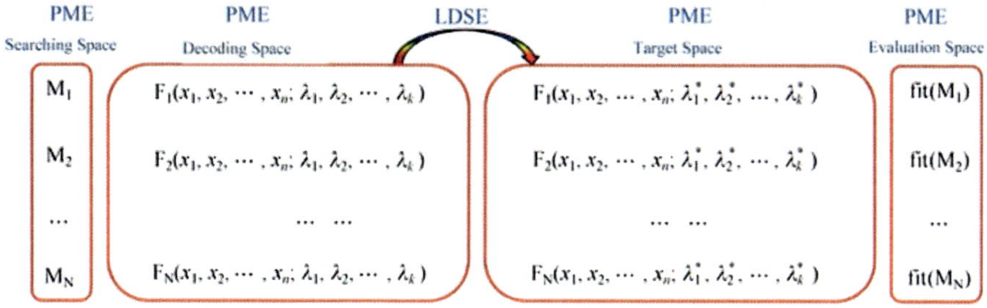

Fig. 4.10 Spatial relationship of PME algorithm

It is worth noting that it does not necessarily reflect the aerodynamic/thermal change trend although the analytical expression obtained by the above algorithm has global optimization characteristics. There are similar counterexamples in traditional interpolation and fitting. For example, polynomial interpolation expression that passes through all points can always be obtained as long as the number of polynomials is increased, in order to cater to the data; spline fitting can also ensure that the model passes all data points, but such a model does not necessarily reflect objective laws. In fact, the improvement of the polynomial coefficients and the segmentation of the spline function raises the complexity of the model, which violates the principle of simplification. That is to say, the complexity of the model must not be too large to obtain a model that can reflect the laws of physics.

We propose a trade-off analysis method to establish a multi-objective optimization model of model correlation and model complexity based on the aforementioned functional optimization algorithm, so as to obtain an analytical function relationship reflecting the aerodynamic thermal change trend under the influence of multiple parameters. It is to find the set of functions with the lowest possible complexity and the highest model accuracy (the degree of agreement with the test data) as high as possible (Fig. 4.11b) with the optimization algorithm from the function space, where each point represents a candidate function, where the red dot is Pareto's most superiority that weighs model complexity and model accuracy, in order to ensure the simplicity of the model function (which can be measured by the complexity of the function), accuracy (which can be measured by the accuracy of the model) and stability (cannot contain singularities, it should have a certain extrapolation ability). Then a correlation model with certain stability is constructed with some functions of the Knee area (lower left corner, yellow ellipse indicated part) of the Pareto solution set.

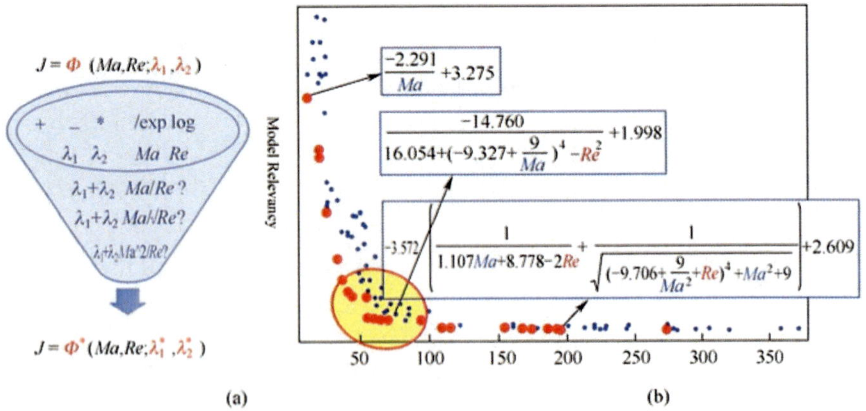

Fig. 4.11 Schematic diagram of the method for excavation of change trend **a** simultaneous optimization of model structure and model optimization; **b** simultaneous optimization of model accuracy and complexity

The method to perform prediction on hypersonic aerodynamic force/heat according to the above method is called the functional prediction method to analyze the change law. This method makes use of a functional intelligent algorithm, in which the input variables are test data and aerodynamic knowledge. The reasoning process does not require manual intervention, which could avoid the influence of the reasoner's subjective factors. The resulting analytic function is globally optimal and conforms to the basic principle of simplicity.

There is another feature of the functional prediction method for analyzing the changing law that the theoretical equation obtained by the simplified analysis of traditional physics can be regarded as a special case of the new method. In fact, any function is a point in space in a functional space, and it is a candidate in the search of functional intelligent algorithm. As a function, the empirical equation is certainly within the scope of functional intelligent algorithm search and evaluation. In essence, the empirical equation can only optimize the coefficient set, but does not optimize the function structure, so it can be regarded as a local optimization. The algorithm of functional intelligent optimization performs intelligent optimization on a higher level of space, which can realize the global optimization of the function structure and coefficient set at the same time (Fig. 4.11a), and achieve complexity control (Fig. 4.11b), so that it is possible to obtain a more comprehensive and objective aerodynamic force/heat change law, in order to provide decision support for the aerodynamic heat world consistency analysis and the prediction of aerodynamics under flight status.

4.3.4 *The Method for Functional Prediction of Invariant Laws*

First, let's review the exploration of invariant laws from previous generation scientists such as Von Karman and Qian Xuesen, etc. Von Karmen used the Lussen number (Kn) as a similar parameter to study the aerodynamic characteristics of the hypersonic rarefied flow field as early as the middle of the twentieth century. Subsequently, Qian Xuesen [22], HK Cheng [23] and others proposed improved parameters, which are proportional to $M_\infty^2/\mathrm{Re}_\infty$, and $C^* M_\infty^2/\mathrm{Re}_\infty$ (see $M_\infty/\sqrt{Re_\infty}$ [24]), where M_∞ is the incoming Mach number, Re_∞ is the incoming Reynolds number, and C* is the correction coefficient. The magic of these parameters is that these parameters can be used to transform them to a smooth curve reflecting invariant laws to varying degrees even with different Mach number test data, and the correlation effect of these three parameters is gradually improving [24] (Fig. 4.12a–c). Under ideal condition, the data distribution tends to be closer to the same curve (Fig. 4.12d).

There are several questions worth considering: ① Is there any possibility for further improvement of these parameters? ② Is the correlation parameter obtained under this dilute flow condition suitable for continuous flow hypersonic aerodynamic

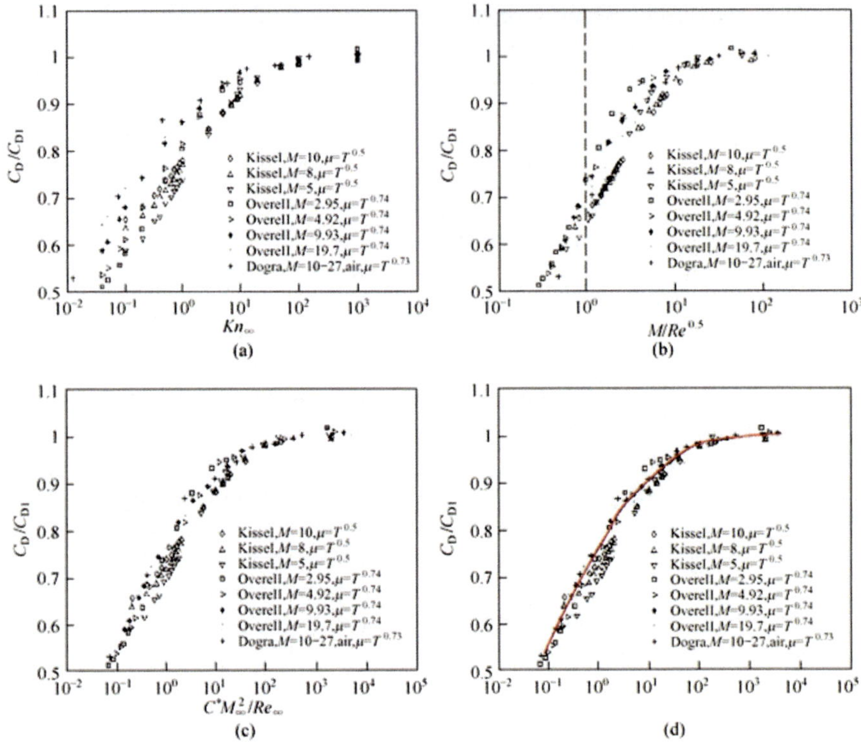

Fig. 4.12 Several early similar parameters

force/heat law? ③ How can we obtain similar parameters if it is no longer applicable? Here are the answers below.

First of all, it can be seen that the similar parameter is essentially a spatial transformation (the similar parameter is its kernel function) from the perspective of functional space from the comparison of several similar parameters in the early stage. Test data is in different degrees of data unification under each different transformation. In other words, the similar parameters from Von Carmen, Qian Xuesen and others are unified in thought and are spatial transformations with different degrees of correlation. Obviously, these similar parameters are just a few special cases of functional space transformation, so there is a possibility on further improvement. In fact, Hozumi et al. have already made relevant attempts [10].

Then, studies have shown that these similar relationships are no longer applicable since the continuous flow hypersonic flow field environment is very different from the dilute gas environment.

Finally, it will focus on how to discover and improve similar correlation parametersconditions in this chapter. In the same way, the acquisition methods of the above similar parameters are all artificial, which requires strong experience and keen insight, and only famous physicists can be competent. Therefore, it is of great significance

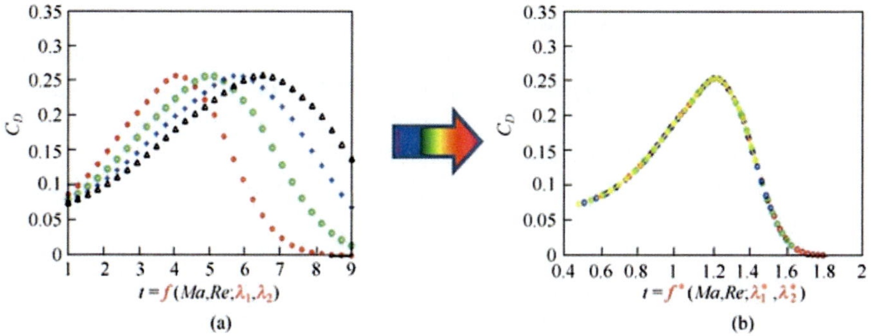

Fig. 4.13 Schematic diagram of adaptive spatial transformation through use of PME **a** data distribution under initial transformation; **b** data distribution under adaptive transformation

if artificial intelligence algorithms can be used to help physicists find the invariant relationships more quickly and obtain the corresponding correlation parameters.

We found and adopted the essence of ideas from Von Karmen, Qian Xuesen and others during their exploration of the invariant law of high-speed thin flow, and proposed an adaptive spatial transformation (AST) method [10], which automatically performs parameter selection and equation optimization, aimed at detection of similar parameters reflecting the constant law of hypersonic flow, so as to predict the flight status more reliably.

AST does not directly fit modeling and prediction with the change trend of the test data, but combine some test state parameters (such as Mach number *Ma*, Reynolds number *Re*, etc.) (such as Von Carmen's $\mathrm{M}_{\infty}\,\mathrm{Re}_{\infty}$, Qian Xuesen's $M_{\infty}/\sqrt{Re_{\infty}}$, Zheng Xian ji's $C^{*}M_{\infty}^{2}/\mathrm{Re}_{\infty}$, etc.), and then examine the results of its comprehensive effect. For example, the test state parameter space is transformed into a one-dimensional subspace through associative parameters (Associative transformations, such as $f(Ma, Re, \lambda_1, \lambda_2)$), and then they are associated with aerodynamic performance parameters in a two-dimensional subspace with them as the horizontal axis of coordinates. In ideal conditions, a correlation curve (Fig. 4.13b) in ideal conditions about a certain test state parameter (Mach number *Ma*). In other words, we expect to get a Variation clusters (unchanged laws) that are "irrelevant" to test state parameters (such as Mach number) by selecting a suitable correlation function (transformation) in ideal conditions.

Or rather, the function curve/surface describes the change trend of function value with the changes of parameters. But it may show different trends (different curves in Fig. 4.13a) often for different parameters (such as different *Ma*, etc.) or different wind tunnels. This cluster of change trends constitutes an invariant law if there is a certain transformation (with f^{*} as the kernel) that can unify the change trends of these parameters (the coincidence curve in Fig. 4.13b), which is not Explicit changes with the change of parameters (such as *Ma*). The kernel function f* is the similar parameter with higher accuracy and reliability that we are looking for.

It is worth noting that "irrelevant" here is not true irrelevant, but refers to that the explicit expression seems irrelevant (or simply explicit irrelevant). For example, the rule curve/surface above is implicitly related to the test state parameter Ma, "irrelevant" is just the result of the combined effect with other parameters. A more intuitive example indicates that as the following:

The transformation law of P with respect to RT for ideal gas should be studied. All the test points are transformed into the subspace PS if 5 $= f(V,T) = T/V$ or any other set of equivalent transformations (such as V/T, (V/T) 2, $\frac{\sqrt{T}}{2\sqrt{T}+\sqrt{V}}$ etc. can be obtained, then these points will form a cluster curve (or straight line cluster), which describes the result of the combined action of V and T. The shape of the curve cluster has nothing to do with V and T explicitly. The discovery of invariant laws through use of comprehensive effects can effectively avoid the coefficient fitting problem of directly using related physical quantities (the reason will not be repeated).

The detection of physical laws through spatial transformation is actually a spatial transformation (taking a point in the functional space) of the original data (some parameters need to be preprocessed), which is projected onto a two-dimensional plane. There is a global optimal transformation under certain conditions. It can also transform them into a curve that reflects a cluster of trends even for data with different parameters (such as Mach number, etc.). It can be called adaptive spatial transformation (AST) because the entire process of finding the optimal transformation kernel function (f^*) is automatically implemented using the functional optimization algorithm PME. It can be concluded the invariant law that the hypersonic aerodynamic performance index meets under different test conditions. The law can be used to predict the data as long as there is no essential change in the physical law. The method can also intuitively see the degree of conformity of the invariant laws and test data in different sections at the same time, so as to give intuitive prediction accuracy under different flight conditions.

4.4 Implementation of the Prediction Method

The realization of the prediction method requires three elements: theory, data and algorithm. First of all, from the perspective of theory, the prediction method is carried out under the guidance of ideas such as multi-space related theory, the thought of "Great law is most simplified" and so on. Secondly, all physical mechanisms under flight conditions must be effectively simulated for the test data set on which the prediction is based, and the test results must be reliable. Therefore, the ability to obtain test data is the key and core factors for the realization of the predictive method, including the wind tunnel test capability, test design level, data standardization and alignment, etc. The prediction method for intelligent optimization based on functional is an algorithm that makes full use of experimental data information under theoretical guidance to discover physical laws. The requirement on algorithm is to be able to achieve global optimization in the functional space. Today, it can be expected that

Table 4.2 Copernicus' planetary motion data (1596)

Planet	Semiaxis	Cycle
Mercury	0.360	0.2408
Venus	0.719	0.6152
Earth	1.000	1.0000
Mars	1.520	1.8809
Jupiter	4.968	11.8621
Saturn	6.716	29.4571

Table 4.3 Tycho's planetary motion data (1619)

Planet	Semiaxis	Cycle
Mercury	0.388	0.2408
Venus	0.724	0.6152
Earth	1	1
Mars	1.524	1.8809
Jupiter	5.2	11.8621
Saturn	9.51	29.4571

the relevant predictive algorithms will definitely make leaps and bounds in the next few years with the rapid development of computing speed and artificial intelligence.

As the following, PME, the intelligent optimization algorithm in the functional space is used to explore the laws of planetary motion data to illustrate the decisive role of data accuracy in the results.

There is the planetary motion data of different years indicated in Tables 4.2, 4.3 and 4.4, of which Copernicus's data accuracy in 1596 is the worst and the data accuracy of Tycho in 1619 has been greatly improved. The data measured by modern means are not much different. The parameters in the table are dimensionless with the long axis R of the earth and the operating period T of 1.0.

Table 4.4 Planetary motion data measured by modern means

Planet	Semiaxis	Cycle
Mercury	0.387099	0.241
Venus	0.723332	0.615
Earth	1	1
Mars	1.523688	1.881
Jupiter	5.202833	11.86
Saturn	9.538762	29.46
Uranus	19.191391	84.32
Neptune	30.061069	164.8
Pluto	39.529402	248.1

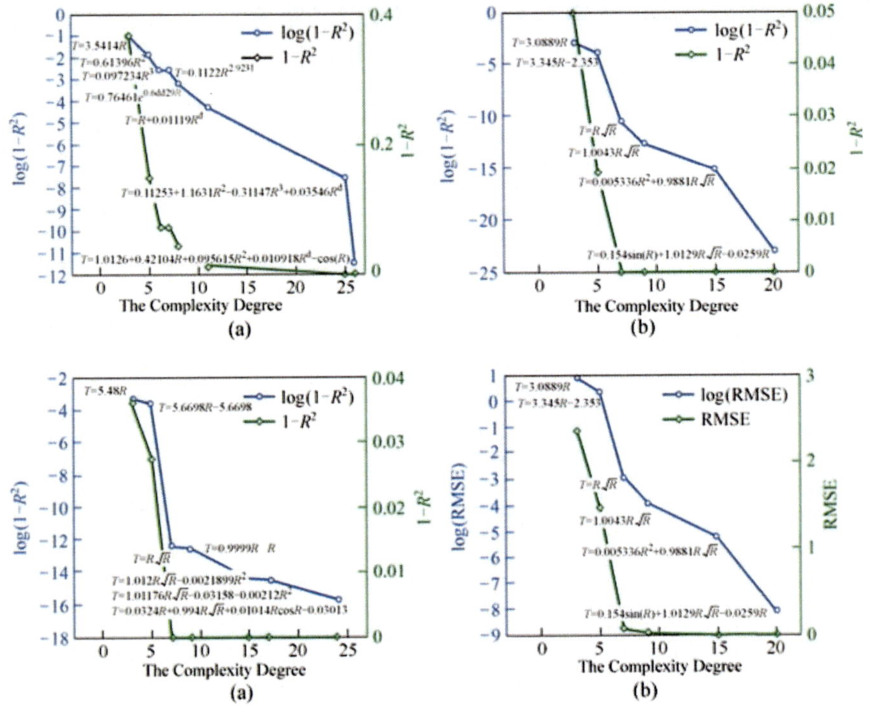

Fig. 4.14 Effect of data accuracy on prediction methods **a** optimization results based on the copernicus data set (1596); **b** Optimization results based on the Tycho data set (1619); **c** Optimization result 1 based on modern data set; **d** optimization result 2 based on modern data set

Their analytical relationship is automatically derived in the functional space using the intelligent optimization algorithm PME, with the period T as the target and the semi-major axis R as the parameter. The results are shown in Fig. 4.14.

As can be seen from Fig. 4.14a, the algorithm cannot find the true physical law in the data from Copernicus in 1596 (the basic form of Kepler's third law $T = R^{3/2}$), the correlation error (Represented by 1-R^2 or log (1-R^2), in which R^2 represents the coefficient of determination) decreases gradually with increasing complexity. It indicates that it cannot get a good rule if the accuracy of the data is not enough, no matter how good the algorithm is. The correlation error index 1-R^2 decreases in a cliff-like manner as the complexity increases if worked on higher-precision observation data from Tycho in 1619 (Fig. 4.14b), and the algorithm can derive the basic form of Kepler's third law $T = RR$.

No matter what kind of correlation error metric is used (such as the previous 1-R^2, log(1-R^2), or root mean square error RMSE, etc.), the algorithm can make the error increase with complexity decreases in a cliff style and reverses the third law of Puppler (Fig. 4.14c, d) if worked on data measured by modern means [26, 27].

This example indicates that the accuracy of the data determines the success or failure of the prediction method. It can be expected that whoever has the precise test data with rich information and the algorithm for global optimization in the functional space will be able to grasp the potential invariant laws in the complex and changing data on the surface.

4.5 Preliminary Application of the Prediction Method

4.5.1 Accuracy Assessment

We used a set of classic test functions to test and evaluate in the seven-dimensional data space, in order to investigate the correlation ability of the multi-dimensional spatial data global optimization correlation method for high-dimensional spatial data. There is the average error of the seven-dimensional spatial data correlation function result in Table 4.5. It can be seen from Table 4.5: The average error of the overall optimization correlation function obtained from 200 supporting basic frame data is 4.92% if there are 5 extreme points in the distribution of aircraft performance parameters in this seven-dimensional data space. The average error of the optimized correlation function is only 1.58% if the aircraft performance parameter distribution has only one extreme point. The accuracy of global optimization correlation functions is quite encouraging compared with traditional correlation algorithms. Generally speaking, different wind tunnels are always used in the development process to continuously carry out tests for an aircraft model. Therefore, the correlation function obtained in this way can make the most of the results of the wind tunnel test, and the more reasonable the correlation rule of the obtained data and the more accurate the correlation result as the wind tunnel test data continues to accumulate. From this point of view, the association method for functional optimization has self-evolution and self-lifting capabilities based on multi-dimensional spatial data correlation theory [28, 29].

Table 4.5 Distribution diagram of prediction methods in seven-dimensional space with the complexity and the number of measurement points

Number of Test Points	1	2	5	10
10	0.0752	0.121	0.124	0.125
20	0.0429	0.0895	0.108	0.109
50	0.0222	0.0541	0.0877	0.0978
100	0.0192	0.0427	0.0603	0.0688
200	0.0158	0.0304	0.0492	0.0528
500	0.0111	0.0224	0.0311	0.0353
1000	0.00735	0.0158	0.0248	0.0279
2000	0.00407	0.0113	0.020	0.0227

Note The number of extreme points in the table represents the complexity of the physical law

4.5.2 Examples of Detection of Aerodynamic Law

The aerodynamic data of two pointed cones with different sizes (length 0.75 and 1.5 m) measured in wind tunnel test (Fig. 4.15) in JF-12, supplemented with more than 2,000 calculation data, and the parameters can cover flight corridor and wind tunnel conditions with Ma 5–9 and altitude of 25–50 km. We obtained the relationship between the world and the earth that can reflect its invariant laws by use of the adaptive spatial transformation method based on functional optimization in this project. The correlation results are shown in Fig. 4.16.

We first obtain the analytical relationship of its axial force coefficient% from the analytical matrix evolution algorithm (PME) for an aircraft based on the obtained test and calculation data within a certain range (the process is omitted):

$$C_A = 0.029 + 0.907(\frac{1}{Ma^2} + \frac{1}{Re^{1/4}} + \frac{0.260T_0}{T_w}) \tag{4.3}$$

The incoming flow Mach number range Ma ∈ [3, 5], the Reynolds number range Re ∈ [3.8 × 10^5, 4.8 × 10^7], and the wall temperature range T_w ∈ [60, 250.3] K.

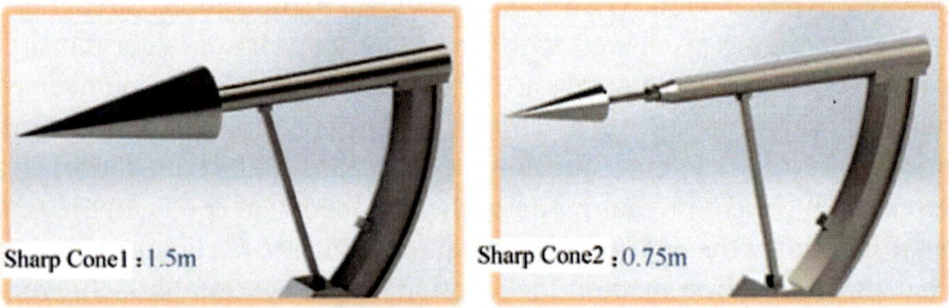

Fig. 4.15 Two wind tunnel test models (including support)

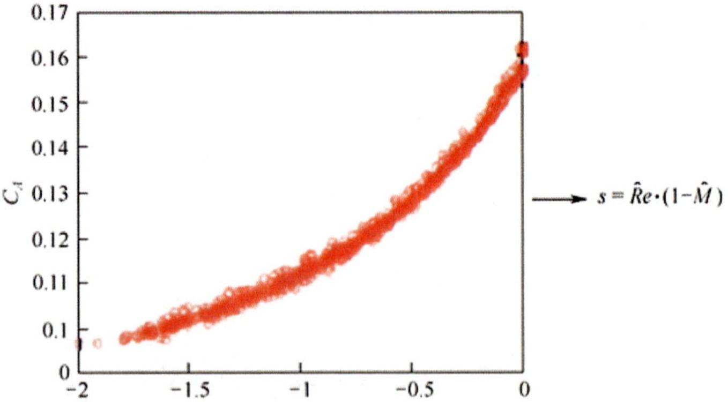

Fig. 4.16 Correlation result of axial force of sharp cone

4.5.3 Examples of Detection of Pneumatic Heat Law

The distribution law of flat plate heat flow data in a test state of JF-12 wind tunnel:

When the angle θ between the plate and the incoming flow is unchanged ($\theta = 10°$), it is set that the heat flow satisfies the relationship: $St = F(x, p, u, H0, Hw, Re_L, Re_x)$, according to the heat flow distribution data along the center of the plate. We get the optimal relationship through the functional space optimization algorithm: $St = k/, \sqrt{Rex}$ where $k = 0.392$.

The heat flow distribution before the correlation is shown in Fig. 4.17 when the angle θ between the plate and the incoming flow changes between [1°, 20°], and the regularity is very poor. Set heat flow satisfy the relationship:

$$St = F(x, \theta, \rho, u, H_0, H_w, R_{eL}, Re_x)$$

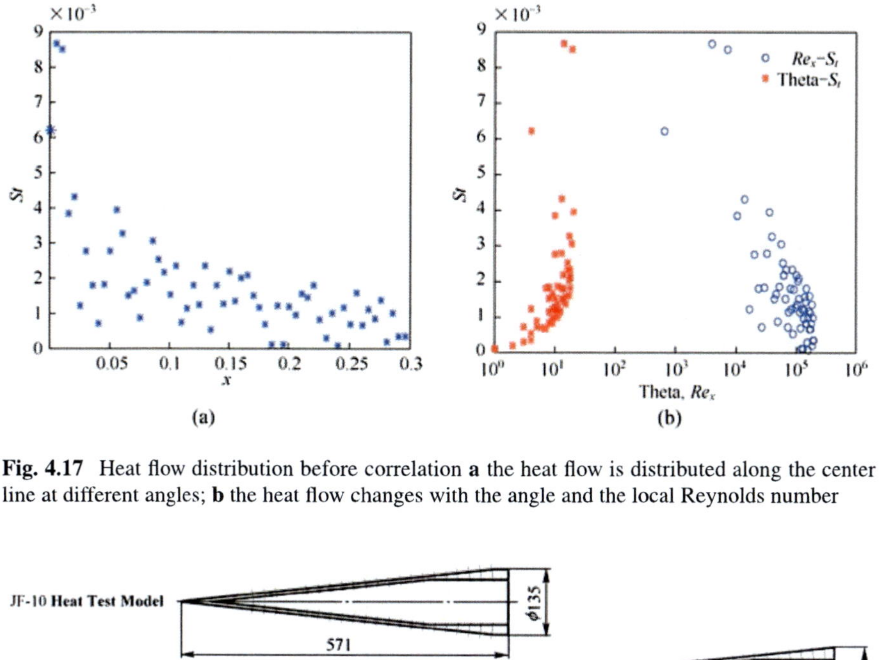

Fig. 4.17 Heat flow distribution before correlation **a** the heat flow is distributed along the center line at different angles; **b** the heat flow changes with the angle and the local Reynolds number

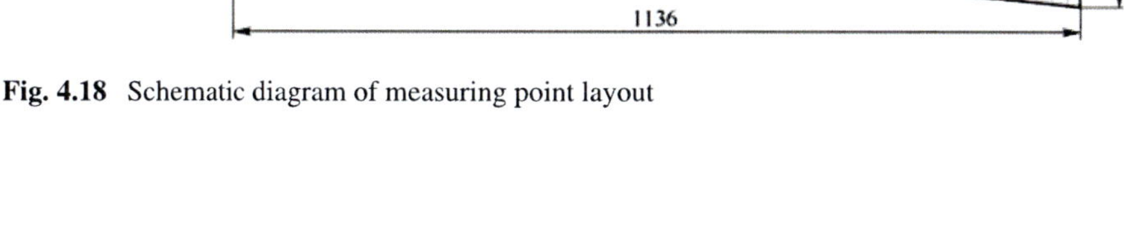

Fig. 4.18 Schematic diagram of measuring point layout

The optimal association relationship is obtained through the functional space optimization algorithm:

$$\mathrm{St} = \mathrm{k} \cdot \sin(\theta) \cdot \sqrt{\cos(\theta)}/\sqrt{Re_x}, \text{ in which, } k = 2.27.$$

The distribution law of heat flow data at the stagnation point of the ball head is as the following:

Let stagnation point heat flow q_{w0} and incoming flow temperature T, pressure P, ball head radius R_n and wall temperature T_w meet the relationship: $q_{w0} = F(M, T, P, R_n, T_w)$, the optimal relationship can be obtained with use of functional space optimization algorithm: $q_{w0} = kM^3\sqrt{\frac{PT}{R_n}}$, where $k = 1.1062$, the units are international standard units (SI).

The detection invariant law of sharp cone heat flow, half cone angle: 7° (Fig. 4.17).

The heat tests were carried out on two different wind tunnels, JF-10 and JF-12, respectively. The test states and the number of test points are as follows: JF-10: $H = 16$ MJ/kg, 4979 m/s, thermochemical non-equilibrium state, $L = 571$ mm, 17 measuring points; JF-12: $H = 3.3$ MJ/kg, 2343 m/s, calorimetric complete gas, $L = 1136$ mm, 37 measuring points.

The arrangement of measuring points is shown in Fig. 4.18, and the measured heat flow is shown in Fig. 4.19a (unit: MW). An adaptive spatial transformation method based on functional optimization is adopted with 11 input parameters: total temperature T_0, total pressure P_0, total enthalpy $H0$, wall enthalpy Hw, incoming pressure P, density ρ, velocity v, Mach number M, temperature T, unit Reynolds number Re_L, local Reynolds number Rex. The algorithm decides to select several parameters to construct the correlation model, the structure and coefficient of the correlation model [30, 31].

The correlation model reflecting the two invariant laws that are commonly observed is obtained after the operation (see Fig. 4.20 for the operation process):

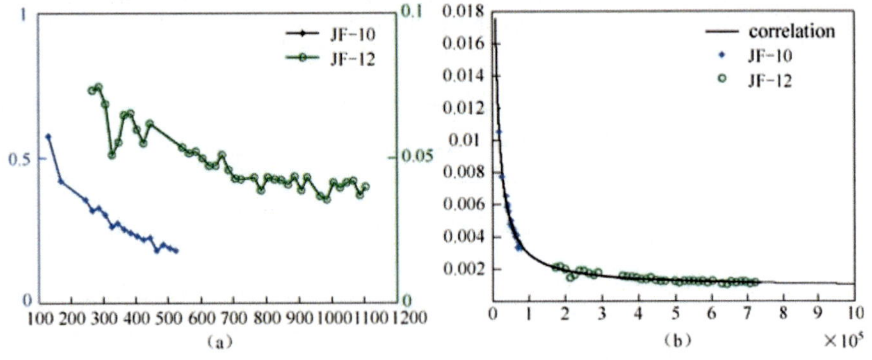

Fig. 4.19 Consistent correlation of data of different wind tunnels **a** the heat flow before the correlation is distributed along the generatrix; **b** the heat flow after the correlation is distributed with the correlation parameters

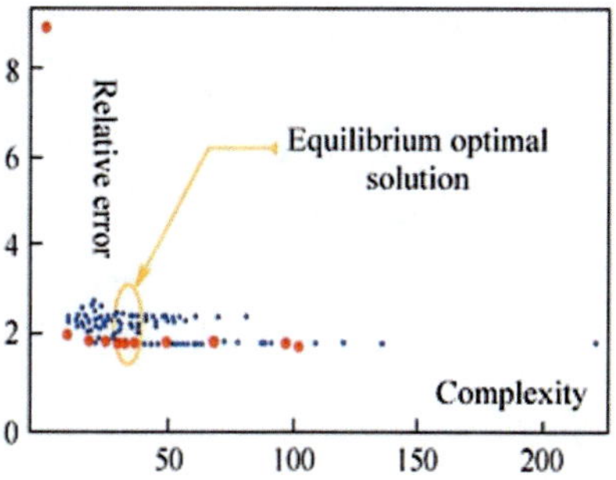

Complexity		$1-R^2$	function
1	17	0.010	$9.18\times10^{-4}+\frac{193.16}{\mathrm{Sqrt}[x_{11}]}$
2	19	0.008	$7.84\times10^{-4}+\frac{231.41}{5363.64+x_{11}}$
3	22	0.008	$7.59\times10^{-4}+\frac{\mathrm{Log}[x_{11}]19.07}{x_{11}}$
4	24	0.008	$8.11\times10^{-4}+\frac{222.10}{x_{11}+\sqrt{x_3}}$
5	31	0.008	$5.52\times10^{-4}+(4.57\times10^{-2})x_6+\frac{238.17}{27.52+x_{11}+x_7}$
6	35	0.008	$5.43\times10^{-4}+(1.31\times10^{-9})x_3^2+\frac{240.94}{269.86+x_{11}+x_7}$
7	38	0.007	$-(1.53\times10^{-3})+\frac{146.99}{x_{11}}-(2.53\times10^{-8})x_{11}+(4.98\times10^{-6})x_{11}x_6+(4.13\times10^{-4})x_8$

Fig. 4.20 Distribution of association transformations in functional space

$$\mathrm{St} = a + \frac{b}{c + Re_x} \tag{4.4}$$

All the coefficients $a = 0.0008223$, $b = 220.1$, $c = 3158$, are dimensionless parameters.

The correlation effect is shown in Fig. 4.19b. It can be seen that each wind tunnel follows its own distribution law before the association, and a very consistent distribution law is obtained after the association. Their experimental flow field parameters and model size are quite different.

As a comparison, the results are given below of the two existing theoretical methods for correlation.

Method one is the classical correlation method $StRe_x^n$ = constant obtained by theoretical simplification, where n = { 0.5Laminar / 0.2Turbulence

It can be concluded whether it is JFT or JF-12 in this case, the flow state is laminar flow, so n takes 0.5.

Another one is the boundary layer outer edge correlation method [25]: $\mathrm{St}_e\sqrt{Re_x}$ = constant.

Interpolation is carried out according to the measured 57 data points and the corresponding models of the two methods are:

$$St\sqrt{Re_x} = 1.1 \tag{4.5}$$

$$St_e\sqrt{Re_x} = 0.73 \tag{4.6}$$

The correlation effect is shown in Fig. 4.21.

It can be seen from Fig. 4.21 that different types of wind tunnel laws can be associated together by using these two existing methods overall, which is much improved compared to before the association (Fig. 4.19a).

However, the so-called secondary factors are ignored since some assumptions need to be made in the process of deriving the correlation equation, which leads

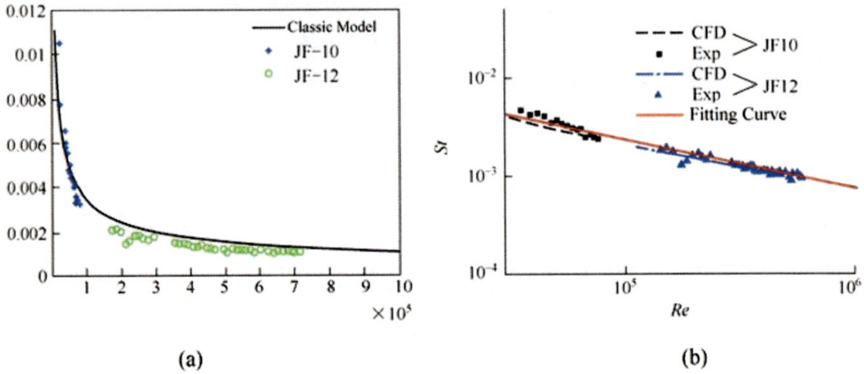

Fig. 4.21 The Association results for existing methods **a** the association results for empirical methods; **b** the association methods for outer boundary layer

to the unnatural connection of different wind tunnel laws. In other words, these equations have not actually been able to truly obtain the invariant laws commonly obeyed that reflect different wind tunnels.

4.6 Summary and Outlook

It mainly introduces an artificial intelligence method for prediction of hypersonic aerodynamic force/heat in this chapter, which is associated prediction method of intelligent optimization based on functional. This method analyzes the problem in a multidimensional space with regarding the flying environment as an ideal wind tunnel, standing at the global height of the wind tunnel group, which breaks through traditional thinking. The analytical variation and invariability law of aerodynamic force/heat of hypersonic flow under certain conditions is explored by using of special functional intelligence optimization algorithm, and this physical law is then used for aerodynamic force/heat prediction. The input variables for this method are test data and aerodynamic knowledge. The reasoning process does not require manual intervention, which could avoid the influence of the reasoner's subjective factors. The resulting analytic function is globally optimal and conforms to the basic principle of simplicity.

Data association and prediction through invariant laws are of higher reliability, however, invariable laws are not universal, that is, the application of invariant laws is also conditional. Roughly speaking, the so-called invariant law obtained from the test data is the invariant law commonly obeyed in different wind tunnels and different test parameters, which reflects different physical mechanisms that the different wind tunnels and different test parameters jointly simulate under hypersonic conditions. It has not been effectively simulated in the ground wind tunnel group if a new physical mechanism (such as high-temperature gas effect, scale effect, etc.) appears

Fig. 4.22 Comparison between the analysis scope and applicable scope of the new methods and the existing methods

under flight conditions, and the influence of this mechanism on the aerodynamic heat environment cannot be ignored. The predicted results obtained by the constant law will no longer be applicable.

In other words, the applicable conditions of the association prediction method of intelligent optimization based on functional include: ① All the physical mechanisms that function under flight conditions are reflected in the wind tunnel simulation. ② The data of wind tunnel is reliable, of course, a certain amount of error is allowed without affecting the simulation of the physical mechanism; ③ no new physical mechanism cannot be ignored under the flight conditions.

More generally, the approximation methods such as fitting/interpolation/regression/neural network/support vector machine that are analyzed in real number space are applicable to the envelope of sample data points, while the association prediction method of intelligent optimization based on functional introduced in this chapter makes use of physical (unchanged) laws for correlation with a wider scope of application as the envelope of the physical mechanism simulated by the wind tunnel group (Fig. 4.22).

References

1. Bushnell DM (2006) Scaling: wind tunnel to flight. Annu Rev Fluid Meeh 38:111–128
2. Engel CD, Praharaj SC (2012) Miniver upgrade for the avid system. Gen Books LLC 1:3
3. Elsenaar A (1988) Onreynolds number effects and simulation. AGARD: Aerodyn Data Accuracy Qual: Requir Capab Wind Tunn Test 22(SEE N 89–16846 09–09)
4. Haines AB (1976) Further evidence and thoughts on scale effects at high subsonic speeds. AGARD Wind Tunn Des Test Tech 12(SEE N 76–25213 16–09)
5. Haines AB (2000) Prediction of scale effects at transonic speeds: current practice and a gaze into the future. Aeronaut J 104(1039):421–431
6. Poisson-Quinton P (1984) Ground flight test techniques and correlation. AGARD Wind Tunn Test Tech 14 p (SEE N 84–23564 14–01)
7. Peterson JJ, Mann M, Sorrells R et al (1980) Wind tunnel flight correlation study of aerodynamic characteristics of a large flexible supersonic cruise airplane (XB - 70 - 1). II—Extrapolation of wind—tunnel data to full scale conditions. NASA TP-1515
8. Morelli EA, Derry SD, Smith MS (2005) Aerodynamic parameter estimation of the X-43A (Hyper-X) from flight test data. AIAA-2005–5921 32:15–19

9. Depriest D, Morgan C (2003) Statistical methods for rapid aerothermal analysis and design technology: validation. NASA Tech Rep, NAG1- 02030(5):66–68
10. Hozumi K, Yamamoto Y, Masson A et al (2001) Study for prediction of flight aerodynamic heating using HYFLEX flight and wind tunnel data. AIAA paper, 2001–1826
11. Lee JH, Kim ET, Chang BH et al (2009) The accuracy of the flight derivative estimates derived from flight data. World Acad Sci Eng Technol 58:843–849
12. Norgaard M, Jorgensen C, Ross J (1997) Neural network prediction of new aircraft design coefficients . NASA Tech Memo 2:112197
13. Rajkumar T, Bardina J (2002) Prediction of aerodynamic coefficients using neural networks for sparse data. FLAIRS-02 Proc 5:242–246
14. Malmathanraj R, Tyson TF (2011) Characteristic prediction of wind tunnel tests using learning from examples . Int J Comput Appl 35(6):5–14
15. Ravikiran N, Ubaidulla P (2004) Support vector machine approach to drag coefficient estimation. In: Proceedings of 7th international conference on signal processing, vol 2, pp 1435–1438
16. Zonglin J, Luo Hu, Zongmin et al (2015) Correlation theory and correlation method of multidimensional space on hypersonic wind tunnel test data. Chin Sci Phys Mech Astron 45:124705
17. Bertin JJ, Cummings RM (2006) Critical hypersonic aerothermo-dynamic phenomena. Ann Rev Fluid Meeh 38:129–157
18. Mark C, Davis J (2008) X-43A Flight-test-determined aerodynamic force and moment characteristics at Mach 7.0. Spacecraft Rockets 45:472–484
19. Koza JR (1992) Genetic programming: on the programming of computers by means of natural selection, 5th edn. MA MIT Press, Cambridge
20. Neill M, Ryan C (2001) Grammatical evolution. IEEE Trans Evol Comput 5:349–358
21. Luo C, Yu B (2012) Low dimensional simplex evolution a new heuristic for global optimization. J Global Optim 52:45–55
22. Tsien HS (1946) Super—aerodynamics, mechanics of rarefied gases . Aerosp Sci 13(2):653–664
23. Cheng HK (1961) Hypersonic shock-layer theory of the stagnation region at low Reynolds number. In: Proceedings of the 1961 heat transfer and fluid mechanics institute, pp 161–175
24. Macrossan MN (2006) Scaling parameters for hypersonic flow: correlation of sphere drag data. In: Proceedings of 25th international symposium on rarefied gas dynamics, vol 12. St. Petersburg: Siberian Branch of the Russian Academy of Sciences, pp 759–764
25. Wang Q, Li JP, Zhao W et al (2016) Comparative study on aerodynamic heating under perfect and nonequilibrium hypersonic flows. China-Phys Mech Astron 59(2):624701
26. Haines AB (1994) Scale effects on aircraft and weapon aerodynamics. AGARD-AG-323, AGARD
27. Luo C, Hu Z, Zhang SL et al (2015) Adaptive space transformation: an invariant based method for predicting aerodynamic coefficients of hypersonic vehicles. Eng Appl Artif Intell 46:93–103
28. Luo C, Zhang SL (2012) Parse-matrix evolution for symbolic regression. Eng Appl Artif Intell 25:1182–1193
29. Rufolo GC, Marini M, Roncioni P et al (2007) In flight aerodynamic experiment for the unmanned space vehicle ftb-1. First CEAS Eur Air Space Conf, Berlin, Germany 2:1–12
30. Rufolo GC, Roncioni P, Marini M et al (2008) Post flight aerodynamic analysis of the experimental vehicle prorausv 1. AIAA 2008–2661(2):16–23
31. Qing W, Weiqi Q, Kaifeng He (2008) Identification and optimized input design on missile aerodynamic parameters . J Astronaut 29(3):789–793

Chapter 5
Numerical Method for Flow Transition Prediction

5.1 Introduction

As a common flow phenomenon in nature, turbulent flow is often accompanied by flow transition. Turbulent/transitional flow exists widely in fluid machinery, which is developed and utilized by people to serve their life in cycle. For example, the boundary layer flow of the wing and fuselage will often change from the laminar state to the turbulent state when various aircraft are on flights, which has a serious impact on the lift and drag of the aircraft and even the aerodynamic heating. The boundary layer flow will also change from laminar flow to turbulent flow for the blades in various rotating machines, which will have impact on the efficiency of the impeller. Therefore, it is very essential for the prediction on turbulent/transitional flow. Researchers have worked out a variety of numerical prediction methods for turbulent/transitional flow at present, including LST stability analysis methods, DNS or LES high-precision finery simulation methods, numerical prediction on turbulent/transitional flow model methods and various engineering experience criteria, etc., which have different predictive capabilities in practical applications. Research will be carried out on these methods to analyze their capabilities.

5.2 e-N Method Based on Stability Theory

Based on the linear stability theory developed in the 1950s, e-N method is the most important method for prediction on natural transition. Widely used in low-speed flow, this method has been widely used in the transition prediction on supersonic flow. It has been used in the transition prediction on hypersonic flow at present and has achieved good performance.

It is an effective method for prediction on turbulence/ transition with technical combination with the e-N method based on the stability theory. This method obtains

M. Zhao, *Prediction and Validation Technologies of Aerodynamic Force and Heat for Hypersonic Vehicle Design*, Springer Aerospace Technology,
https://doi.org/10.1007/978-981-33-6526-1_5

the disturbance growth at first based on the linear stability theory (LST) or solving the parabolic stability equation (PSE), and then obtains the integral of the disturbance growth rate at a specific frequency with the e-N method to predict the flow transition. The e-N method has been widely used at low speeds based on the linear stability theory LST, which is gradually extended to hypersonic flow.

Established based on LST, e-N method is the most famous semi-empirical prediction method for boundary layer transition, which was developed by SmithE and van Ingen [1–3], respectively. Through this method, the envelope of the growth curve is obtained by examining the growth of the amplitude of the disturbance wave at different frequencies, and the growth multiple on the envelope can be obtained, that is, the value of N, of which the position where the transition starts (or the Reynolds number) is where it reaches the preset value (or Reynolds number). This method can be expressed by the following equation

$$N = ln(A/A_0)_{max} \tag{5.1}$$

In the equation: N is the cumulative growth rate of disturbance. A is the amplitude of the disturbance wave after growth, of which the unit is consistent with the selected disturbance physical quantity. A_0 is the initial disturbance wave amplitude, of which the unit is consistent with the selected disturbance physical quantity.

Smith recommends N to be 9 and van Ingen recommends N to be 7 or 8 after comparing with a large number of low-speed wind tunnel test data, which is used to judge the transition position and achieve good results.

e-N method gets rid of the constraint of solving the amplitude of the disturbance wave, and in turn it solves for the ratio of the amplitude, which greatly reduces the difficulty of calculation, so it is widely used.

Its scope of application is limited since this method is based on a series of premises. Firstly, the method is based on the linear stability theory, while both of the basic assumptions of the linear stability theory are small disturbances and parallel flow, which limits the method to be adopted only to research on the natural transition process caused by small disturbances. And the target flow should not be too complicated, it must be a boundary layer flow with parallel (or nearly parallel) streamlines. In addition, the description of the linear stability theory is for the process of disturbing linear growth, which does not cover all the processes of natural transition. The process of natural transition also includes the non-linear growth of susceptibility and perturbation and both of the processes cannot be considered by the e-N method. The influence of these factors needs to be reflected by the "semi-empirical" method just because of this, that is, the transition can be judged by selecting the adjusted value of N, which makes the value of N not universal. There is no theoretical guidance for the selection of N value for a new flow, which must be determined by wind tunnel test or flight test data. However, the *e-N* method can still provide valuable information for transition prediction as a theoretical method based on the physical process of natural transition.

In recent years, the United States has developed STABL software 35 in order to meet the requirements of hypersonic boundary layer stability and transition analysis.

It is believed by Americans that the combination of PSE and e-N method is more promising than the LST-based e-N method to deliver better results consistent with the experimental data. Therefore, the PSE-Chem program is embedded in the STABL software, which solves the parabolic stability equation, and can also take the finite chemical reaction rate and translation-vibration energy conversion mechanism into consideration.

The specific process for the implementation of the e-N transition prediction method based on PSE is that the spatial evolution process of the specific frequency disturbance wave is first obtained through the PSE-Chem analysis program in the boundary layer under the two-dimensional or axisymmetric shape, and then the integrals for growth rate is obtained for the specific frequency disturbance through the e-N method. The integrals are defined as:

$$N(\omega, s) = \int_{s_0}^{s} \sigma ds \tag{5.2}$$

In this equation:

N—Cumulative Growth Rate of Disturbance;

ω—Disturbance Frequency (s^{-1});

s—Range of Disturbance Propagation along the Surface (m);

s_0—Initial Growth Position of Disturbance;

σ—Disturbance Growth Rate, defined as:

$$\sigma = -Im(\alpha) + \frac{1}{2E}\frac{dE}{ds} \tag{5.3}$$

In this equation:

α—Complex Flow Wave Number;

E—Disturbance Kinetic Energy (kg/(ms^2)). It is defined as:

$$E = \int_{n} \overline{\rho}\left(|u'|^2 + |v'|^2 + |w'|^2\right) dn \tag{5.4}$$

In this equation:

ρ—Average Density (kg/m^3);

u', v', w'—Complex Pulsating Velocity Component (m/s^2);

n—Normal Direction of Surface.

The United States applied this method to the research in the HIFiRE project. It is indicated in the results that the N value corresponding to the transition of the flow on different smooth surfaces is about 8–11 for the supersonic conical boundary layer flow, under the condition of small disturbance free flow. The growth of surface roughness and incoming noise level will result in a small selection of N value, which can provide a transition starting position consistent with the conventional wind tunnel measurement when the N value is about 5.5. Johnson and Candler [4] applied this method to analyze the transition problem of hypersonic cone boundary layer in 2006. The model includes two cones with a half cone angle of 5° and 22° respectively, and the Mach number of the coming flow is 20–22. The latest development of STABL software, especially PSE-Chem, is demonstrated through the analysis of these two working conditions. This software can also easily calculate the stability of other operating conditions with these developments. It is believed in researches that other flight test data should be continuously analyzed and the boundary layer transition prediction method combined with PSE and e-N methods should be continuously developed. The PSE-Chem method should continue to be used to evaluate the importance of these factors on the transition for some influencing factors that were not considered in the test, such as head ablation, angle of attack, and heat deformation of the aircraft body.

In addition, there is a research on the effect of incoming turbulence on the N value. Mack [5] fitted the relationship between the inflow turbulence Tu and the N value according to the experimental data from Dryden [6], which can be regarded as a modification of the e-N method, but only applicable to low-speed plate boundary layer flow. And the scope of incoming turbulence is $10^{-3} < Tu < 10^{-2}$.

$$N = -8.43 - 2.4 lnT_u \tag{5.5}$$

In this equation:

N—Cumulative Growth Rate of Disturbance;

T_u—Turbulivity of Incoming Flow.

In China, Su Caihong and Zhou Hengkou from Tianjin University carried out many years of research. The e-N method was improved in 2009, and the transition prediction on hypersonic cone boundary layer was carried out. The flow is significantly improved around the cone with an angle of attack of 1° compared with the traditional e-N method, which is more consistent with the experimental measurement data. The change in the N value of the disturbance wave starting at position 0.85 with x. The transition lines obtained through the improved e-N method, which are the transition line calculated based on the basic flow obtained by the boundary layer equation and the transition line calculated based on the basic flow obtained by direct numerical simulation, respectively.

Luo Jisheng [7] also carried out a lot of work in the research on the improvement of e-N method, studied the boundary layer transition mechanism and susceptibility characteristics, and realized the hypersonic three-dimensional boundary layer e-N method transition prediction. It is shown in Fig. 5.1 that the distribution of the N value predicted by the e-N method, and the comparison of corresponding predicted N values and test data is indicated in Fig. 5.2. The test data is from the cone wind test from the Schneider team at the Purdue University in the United States.

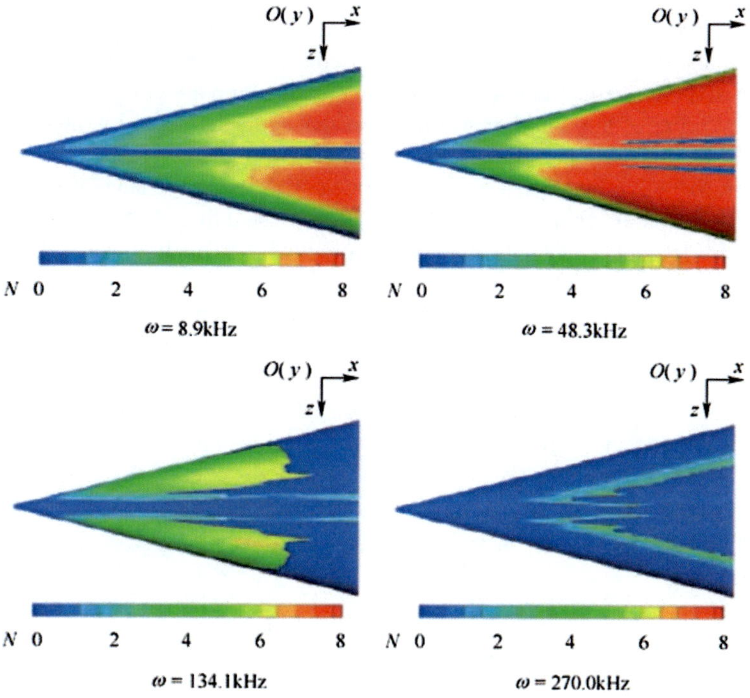

Fig. 5.1 N-value Distribution of Disturbing Waves for Different Frequencies [7]

Fig. 5.2 Comparison of N-value Contour Line and Test Results [7]

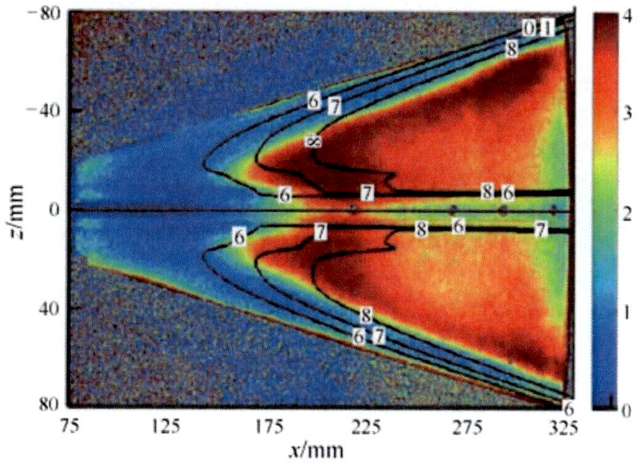

It is show in the research that the previous e-N method is based on the Neutral Curve to integrate the growth rate of the disturbance. The points on the neutral curve are the instability points for the two-dimensional disturbance, therefore the traditional neutral curve refers to the neutral curve of the two-dimensional disturbance. However, the three-dimensional disturbance for the boundary layer flow occurs instability in more forward positions. The neutral curve corresponding to the three-dimensional disturbance is called Zarf curve. An improved method is proposed based on this. The improved e-N method changes the starting point of the integration at first, from the neutral point of the two-dimensional disturbance to the neutral point of the three-dimensional disturbance, that is, the integration from the neutral curve is changed to that from the Zarf curve. Then, the integrated growth rate is changed. The integrated growth rate is the growth rate of the two-dimensional disturbance corresponding to a certain frequency at each station in the previous e-N method, while the improved e-^{N} method selects each station the maximum growth rate corresponding to a certain frequency, and this growth rate is often in three dimensions. The change in the growth rate of disturbance waves with frequency for different booths in a hypersonic inlet area is shown in Fig. 5.3. The neutral curve and Zarf curve are shown in Fig. 5.4, and the integral obtained from different curves. N value is shown in Fig. 5.5, from which it can be seen that the growth rate of three-dimensional disturbance is still very large.

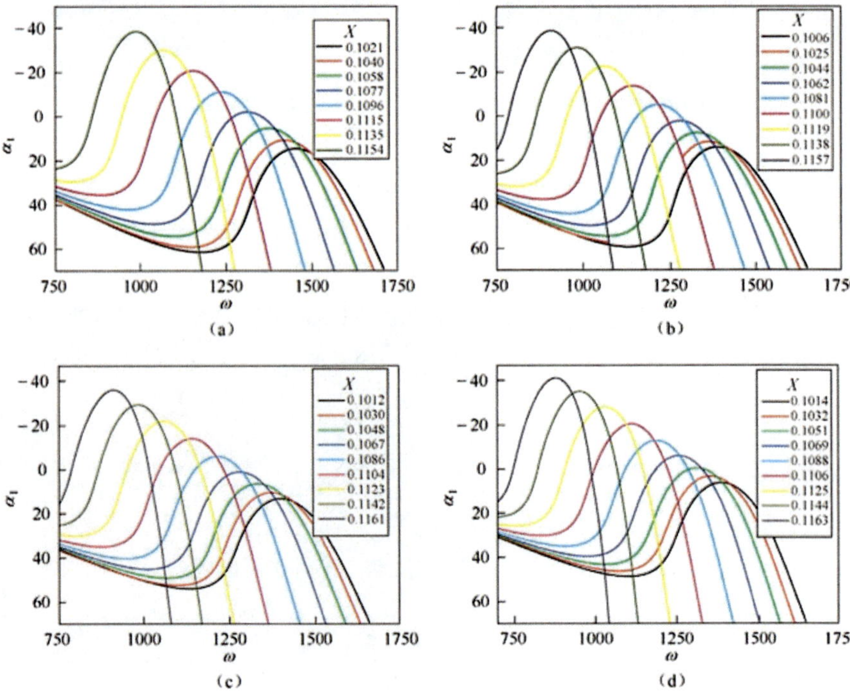

Fig. 5.3 Change of Disturbance Growth Rate with Frequency for Different Display Positions. **a** Z ≈ 0.0; **b** Z ≈ 0. 006; **c** ≈ 0.0125; **d** 7 ≈ 0.0185

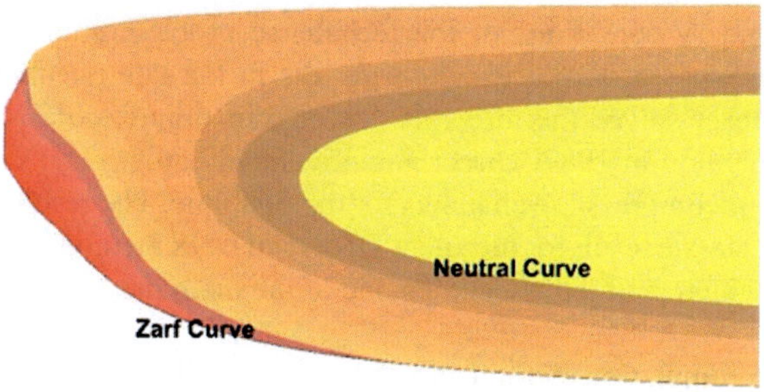

Fig. 5.4 Neutral Curve and Zarf Curve

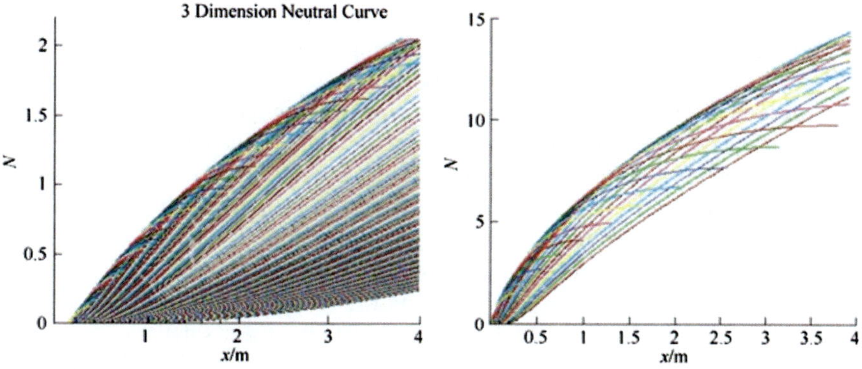

Fig. 5.5 The N Value Integrated from the Neutral Curve and the Zarf Curve

5.3 High-Fidelity LES or DNS Simulation Method

The basic idea of the large eddy simulation (LES) method is to decompose the pulsating motion with the filtering method in the turbulent field into large-scale pulsation and small-scale pulsation, and directly calculate the large-scale pulsation, while the influence of small-scale pulsation on large-scale pulsation is characterized by appropriate model. The idea of LES was first proposed by Smagorinsky [8] and adopted in meteorological research. The development and testing of large eddy simulation ideas in early stages mainly focused on isotropic turbulence [9] and fully development of channel flow [10–12]. The progress of research in the LES method mainly focuses on the sub-grid model. There are related models including Smagorinsky model [8], gradient model [13], similarity model [14], dynamic eddy-viscosity model [15], dynamic mixed mod-el [16], dynamic Clark model [17]. There are also some other models in addition to above.

Direct Numerical Simulation (DNS) is to solve the complete three-dimensional N-S equation directly, to give the evolution process of the characteristic physical quantities in a wide scale range in time and space in the turbulent flow field combined with the boundary conditions of the turbulence problem studied. Researches for Turbulent DNS has achieved excellent results in the late twentieth century. Kim, Moin, and Moser solved the three-dimensional incompressible N-S equation with the spectral method in 1987. Direct numerical simulation results of turbulent flow were given in incompressible channels for the first time. The result of the calculation is a relatively classic result for turbulent DNS, and DNS method is widely recognized by scholars, which gradually becomes one of the main means to study the turbulent mechanism.

Large eddy simulation method and direct numerical simulation method should be combined with high-precision calculation method. Under normal conditions, the non-viscous term for spatial dispersion adopts the fifth-order accuracy WENO format, the fourth-order skew-symmetric form (skew-symmetric form) central format or the higher-order compact format. The viscous term adopts the sixth-order central difference. Temporal dispersion adopts the Runge–Kutta method with the third-order TVD property, and selects appropriate boundary conditions for the wall.

Mesoscopic flow to capture turbulent flow is required in the direct numerical simulation and large eddy simulation methods, and the calculation amount is quite large, which limits the scope to use both of the methods and is only applicable to the local range of flow simulation. DNS method is used to calculate the whole turbulent energy spectrum directly, which is the most accurate method for calculating turbulence. However, in theory, the amount of DNS grid is Re9/4, and the requirement for grid is $10 \leq \Delta x^{+} \leq 20$, $\Delta y^{+}{}_{\text{wall}} < 1$, $5 \leq \Delta z^{+} \leq 10$, which greatly limits the engineering application of DNS, so as to obtain correct simulation results. The requirements from the LES method on the grid are similar to the DNS near the wall surface. The requirements on the grid in other areas are more restrictive than that for the DNS. The requirements for the grid are $50 \leq \Delta x^{+} \leq 150$, $\Delta y^{+}{}_{\text{wall}} < 1{,}15 \leq \Delta z^{+} \leq 40$ and the grid volume can be greatly reduced. In addition, its large-scale pulsation is directly calculated, so the pulsation volume can be calculated. However, the grid volume is also a major limitation in the application of the LES method in a large size range.

External disturbance sources need to be added if considering the influence of external noise environment and turbulence When using LES method and DNS method to study turbulence/transition. However, the external environment is always very complicated, and it is difficult to express in mathematical form, therefore it is difficult to reflect the external environment in truth. That is only can be done by adding some known mathematical forms of disturbances to mimic external disturbances. The methods of LES and DNS are currently have achieved good results in the research on the development of external disturbances and sensitivity problems. There are three forms of external disturbances that are common in nature, which are banding, acoustic disturbances and gust disturbances. The interaction processes between the evolved TS disturbance waves are all different with completely different characteristics. However, the LES method and the DNS method still cannot simulate interference in real external environment.

The LES method and the DNS method are most widely adopted in the field of low-speed flow, and there are problems on computational stability in the field of supersonic and hypersonic flows. The LES method and the DNS method require simultaneous accurate capture of shock waves and accurate description of flow details when applied to hypersonic flow, which carries difficulties to the calculation method. There are mixed format and higher-order WENO formats in high precision numerical method used by LES method and DNS method in hypersonic flow simulation. The mixed format adopts the welcome style in the discontinuous area, and adopts the central differential format or the compact format in the continuous area, in which the weighting is used of the two formats. The format is in high resolution in the smooth area, but it is easy to cause non-physical oscillations during format conversion. There is possibility to get un-physical solution when solving it. The higher-order WENO format is established based on the idea of template weighting. The flow of the entire field is solved through a unified format with high accuracy. There are many improved WENO formats at present, such as the WENO-Z method and the ESWENO method.

The spatial filtering function is adopted in the LES method to perform the Favre filtering operation on the three-dimensional compressible Navier–Stokes equations, all using the International System of Units, defining Favre filtering $\tilde{\varphi} = \overline{\rho\varphi}/\overline{\rho}$, $\overline{\varphi}$ represents the spatial filtering variable. The filtered N-S equations can be written as:

$$\frac{\partial \overline{\rho}}{\partial t} + \frac{\partial (\overline{\rho}\tilde{u}_i)}{\partial x_i} = 0 \tag{5.6}$$

$$\frac{\partial (\overline{\rho}\tilde{u}_i)}{\partial t} + \frac{\partial \left(\rho\tilde{u}_i\tilde{u}_j\right)}{\partial x_j} = -\frac{\partial \overline{p}}{\partial x_i} + \frac{\partial \left(\tilde{\tau}_{ij} - \tau_{ij}^{SGS} + D_{ij}^{SGS}\right)}{\partial x_j} \tag{5.7}$$

$$\frac{\partial \left(\overline{\rho}\tilde{E}\right)}{\partial t} + \frac{\partial \left(\overline{\rho}\overline{E} + \overline{p}\right)\tilde{u}_i}{\partial x_i} = \frac{\partial}{\partial x_j}\left(-\tilde{q}_i + \tilde{u}_j\tilde{\tau}_{ij} + \psi_i^{SGS} + \sigma_i^{SGS} - Q_i^{SGS} - H_i^{SGS}\right) \tag{5.8}$$

In the equation:

ρ—Density (kg/m^3);

u—Speed (m/s);

p—Pressure (Pa);

τ—Tensor of Viscous Stress (N/m2);

E—Energy Item (J/kg);

q—Heat Flux (J/(m^2·s));

t—time (s);

x—coordinate (m);

$\tau_{ij}{}^{SGS}$—SGS Diffusion Flux;

D^{ijSGS}—Nonlinear Item of SGS Viscous Stress;

ψ_i^{SGS}—Diffusion Term of SGS Turbulent;

σ_i^{SGS}—SGS Viscous Diffusion Term;

Q_i^{SGS}—Nonlinear Term of Heat Flux;

H_i^{SGS}—SGS Energy Term;

i, *j*—Subscript, Dumb Symbol;

SGS—Subscript, Sub-grid Item.
The tensor of viscous stress and heat flux are respectively

$$\tilde{\tau}_{ij} = \tilde{\mu}\left(2\tilde{S}_{ij} - \frac{2}{3}\delta_{ij}\tilde{S}_{kk}\right) \tag{5.9}$$

$$\tilde{q}_i = -\frac{\tilde{\mu}C_p}{Pr\frac{\partial \tilde{T}}{\partial x_i}} \tag{5.10}$$

In this equation:

μ—Viscosity Coefficient (N·s/m^2);
*S*stress tensor (s^{-1}), $S_{ij} = (\partial u_i/\partial x_j + \partial u_j/\partial x_i)/2$;

Cp—Specific Heat at Constant Pressure (J/(kg·K));

T—Temperature (K);

Pr—Prandtl Number.
The sub-grid to be sealed is defined as:

$$\tau_{ij}^{SGS} = \overline{\rho}\left(\frac{\overline{\rho u_i u_j}}{\overline{\rho}} - \tilde{u}_i\tilde{u}_j\right) \tag{5.11}$$

$$D_{ij}^{SGS} = \left(\overline{\tau}_{ij} - \tilde{\tau}_{ij}\right) \tag{5.12}$$

$$\psi_i^{SGS} = -\frac{1}{2}\overline{\rho}\left(\frac{\overline{\rho u_i u_k u_k}}{\overline{\rho}} - \tilde{u}_i\frac{\overline{\rho u_k u_k}}{\overline{\rho}}\right) \tag{5.13}$$

$$\sigma_i^{SGS} = \left(\overline{u_j\tau_{ij}} - \tilde{u}_j\tilde{\tau}_{ij}\right) \tag{5.14}$$

$$H_i^{SGS} = C_p\left(\overline{\rho}\frac{\overline{\rho u_i T}}{\overline{\rho}} - \overline{\rho}\tilde{u}_i\tilde{T}\right) \tag{5.15}$$

$$Q_i^{SGS} = \overline{q}_i - \tilde{q}_i \tag{5.16}$$

In this equation:

subscript *i, j, k*—Dumb Symbol.

Among the above unclosed items, the viscous stress nonlinear item D_{ij}^{SGS} and the sub-grid viscous dissipation term σ_i^{SGS} can be neglected due to their small contribution to the energy equation. The remaining items are approximated by establishing the sub-grid model viscosity coefficient μ_t to make the equation closed, and the form is as follows:

$$Q_i^{SGS} = -\frac{\widetilde{\mu}_t C_p}{Pr \frac{\partial \widetilde{T}}{\partial x_i}} \tag{5.17}$$

$$\widetilde{\tau}_{ij}^{SGS} = \widetilde{\mu}_t\left(2\widetilde{S}_{ij} - \frac{2}{3}\delta_{ij}\widetilde{S}_{kk}\right) \tag{5.18}$$

$$H_i^{SGS} = -\frac{\widetilde{\mu}_t C_p}{Pr_t}\frac{\partial \widetilde{T}}{\partial x_i} \tag{5.19}$$

$$\psi_i^{SGS} = \widetilde{u}_j \tau_{ij}^{SGS} \tag{5.20}$$

Zhu Zhibin et al. [18] from China Academy of Aerospace Aerodynamics developed a high-order compact format with DNS numerical simulation methods and carried out research on transition problems caused by cylindrical rough elements. The calculation results of the heat flux on the object surface is shown in Fig. 5.6. The infrared thermal imaging test results is shown in Fig. 5.7. The calculated vortex structure of the flow field is given in Fig. 5.8. In Fig. 5.9, LES numerical simulation method is adopted to calculate the vortex structure of transition flow field caused by several square column rough elements.

Fang Jian et al. from Beijing University of Aeronautics and Astronautics [19] carried out research with large eddy simulation on the effect of wall-oriented high-frequency vibration on the turbulence of low Reynolds number compressible channels, and investigated the effects of turbulent coherent structures on the temperature field and heat transport in the boundary layer. It was found that the velocity coherent structure and the temperature structure in the channel flow always maintain a high degree of consistency. The turbulence in the boundary layer decreases the momentum

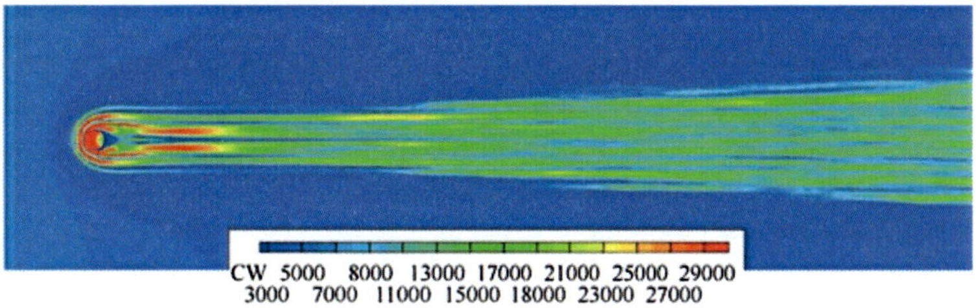

Fig. 5.6 Calculation results of heat flux on the object surface [18]

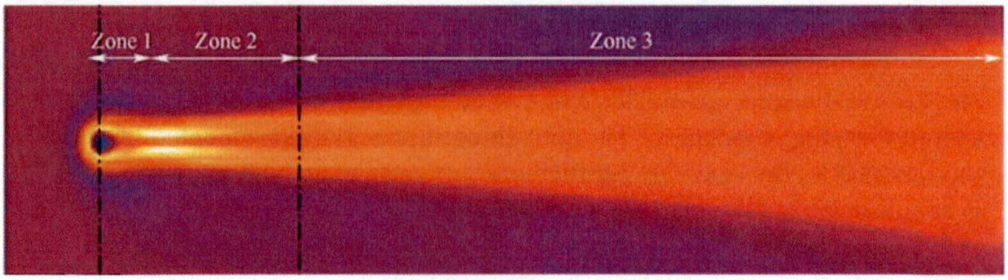

Fig. 5.7 Results of infrared heat imaging test [18]

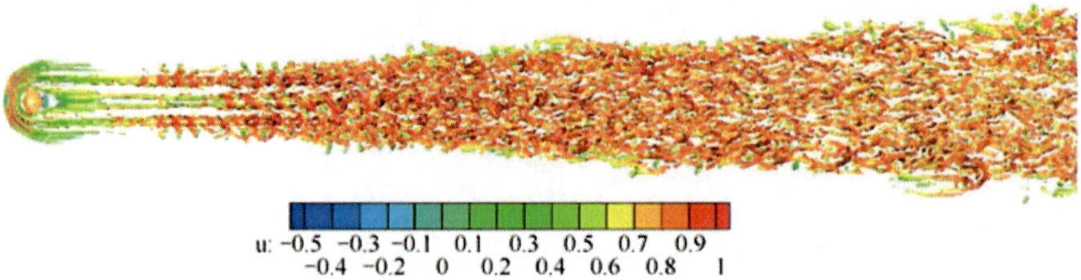

Fig. 5.8 The vortex structure of the flow field (Q Criterion, Flow Velocity Coloring) [18]

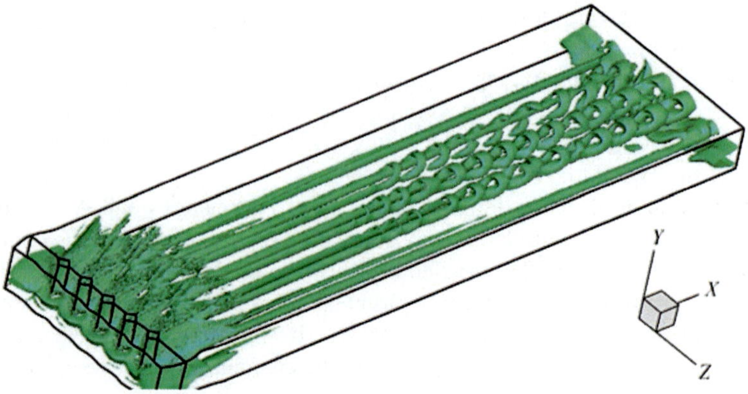

Fig. 5.9 Vortex structure of transition flow field caused by rough element

transport and the heat transport at the same time when the velocity coherent structure is suppressed, and the average wall heat flux is reduced. The evolution of the coherent structure represented by the iso surface of λ_2. The generation of the coherent structure is effectively suppressed as the vibration progresses. The coherent structure almost disappears at the minimum stress (T^+ ~825), and the flow is close to the state of laminar fluidization.

5.4 Turbulence and Transition Model

The numerical prediction method of transition model is a very effective method to solve the transition problem for a long period of time from now to the future. It can master the characteristics of transition flow by establishing a suitable theoretical model with low requirements for calculation conditions and short calculation period. It is not only used for research on turbulence/transition, but also suitable for practical engineering applications. The transition prediction method is more applicable and reliable for the establishment of transport equation of the transition model.

The turbulence models used for transition studies at early stage are mainly algebraic models and single-equation models. Typical ones are the algebraic models from Baldwin and Lomax [20], single-equation models from McDonald and Fish [21], and Arad [22]. However, these models are not available to predict the starting position of transitions on objects with complex shapes, which must be given in advance. Therefore, the development of prediction methods for transitional turbulence model is very promising.

Low Reynolds number transition model is usually used to solve turbulent transition problems with wall constraints in engineering practice. This method can not only simulate the flow in the near-wall region dominated by the viscous effect, but also apply to the fully developed turbulent flow with wider applicability. Priddin [23] conducted the earliest research on low Reynolds number model, and then Scheuerer [24] proposed a low Reynolds number k-ε model with transition prediction ability.

Craft [25] et al. proposed a third-order nonlinear eddy viscosity model to simulate the transition process of the low-velocity plate boundary layer, considering the anisotropy of turbulence and the non-local equilibrium effect. Hadzic [26] also developed a second-order moment model of low-Reynolds number to research on transitions on turbine blades and transitions caused by separated flows. Wilcox [27] also carried out research on transition and established a low Reynolds number k-model that can simulate transition processes. Schmidt [28] and Johnson [29] also improved this kind of method, and restricted the kinetic energy generation term Pk of turbulent flow, but the effect is not obvious. Afterwards, Schmidt and Patankar [30] proposed a correction method for this turbulent kinetic energy generation term, which can extend the development length of the transition zone, but the scope of application is still limited. The low Reynolds number model is in a lack of the structure of the transition mechanism, which needs to be developed to a large extent.

The flow of laminar and turbulent flows will alternately change at the same spatial position in the process of flow transition, which is called the intermittent phenomenon. Starting from this, people have developed new transition models by constructing intermittent factors. Libby [31] proposed the above idea at first, and Dopazo [32] perfected it. Steelant and Dick [33] proposed the transport equation of intermittent factor and simulated the transition of the boundary layer. The calculation results are in good agreement with the experimental data in the direction of flow. Huang and Suzen [34] improved this model, and established a new intermittent factor transport equation with combination of the Cho and Chung models. The calculation results are in better agreement with the experiment. Therefore, the transition model incorporating the intermittent factor transport equation has been greatly developed in the turbulence model. Bradshaw [35] found that pressure pulsation is the main cause for the disturbance of incoming flow into the boundary layer and continuous evolution, so Walters and Leylek [36] and Volino [37] proposed a transition model based on local variables. Menter [38] deliver the equation of intermittent factor and equation of momentum thickness Reynolds number. Fu Song et al. [39] also established a k-m-y model, and there are also many researchers who have accomplished a lot of work. The numerical prediction methods of these two transition models will be mainly introduced here since the practical application effect of Menter transition model and Fusong transition model is very good.

First, Langtry and Menter [40] improved the γ-*Re* correlation transition model, who introduced the transport equation of γ intermittent factor and established the Transport Equation of Re Momentum Thickness Reynolds Number. Two empirical relations between the length of transition region in this model and the critical Reynolds number are established through the analysis of a large number of experimental data, which makes the model reliable for transition prediction. The parameters in this model depend on local variables, which are compatible with modern CFD technology and also suitable for unstructured grids and large-scale parallel computing. The new transition model with application of local variables has been applied in a considerable number of civil aircraft flow simulations, which has achieved some results in the prediction of the transition of the aircraft with a hypersonic Mach number below 8.

The γ intermittent factor transport equation controls the generation of transition by controlling the generation of turbulent kinetic energy in the boundary layer in the γ-*Re* model. The Re transport equation is adopted to calculate the local transition Reynolds number, which can reflect the incoming turbulence and the influence of pressure gradient in the boundary layer.

The transport equation of intermittent factor is as the following:

$$\frac{\partial(\rho\gamma)}{\partial t}+\frac{\partial\left(\rho U_j\gamma\right)}{\partial x_j}=P_\gamma-E_\gamma+\frac{\partial}{\partial x_j}\left[\left(\mu+\frac{\mu_t}{\sigma_f}\right)\frac{\partial\gamma}{\partial x_j}\right] \tag{5.21}$$

In this equation:

ρ—Density (kg/m^3);

$U_{\rm j}$—Speed in j Direction (m/s);

γ—Intermittent Factor;

P_γ—Generation Item of Intermittent Factor Equation;

E_γ—the Dissipative Item of the Intermittent Factor Equation;

μ—Coefficient of Viscosity (kg/(m·s));

μ_∞—Viscosity Coefficient of Turbulent Flow (kg / (m-s));

x_j—Coordinate in the Direction of j;

t—Time (s);

σ_f—Model Constant, 1.0.

The equation for momentum thickness Reynolds number Re_θ

$$\frac{\partial\left(\rho\overline{Re_{\theta t}}\right)}{\partial t}+\frac{\partial\left(\rho U_j\overline{Re_{\theta t}}\right)}{\partial x_j}=P_{\theta t}+\frac{\partial}{\partial x_j}\left[\sigma_{\theta t}(\mu+\mu_t)\frac{\partial\overline{Re_{\theta t}}}{\partial x_j}\right] \tag{5.22}$$

In this equation:

ρ—Density (kg/m^3);

U_j—Speed in j Direction (m/s);

$\overline{Re_{\theta t}}$—Reynolds number of momentum thickness at the beginning of local transition;

$P_{\theta t}$—the source item of the momentum thickness Reynolds number equation;

μ—Viscosity Coefficient (kg / (m-s));

μ_t—Viscosity Coefficient of Turbulent Flow (kg / (m·s));

x_j—Coordinate in the Direction of j;

t—Time (s);

$\sigma_{\theta t}$—Model constant, 2.0.

The transition length F_{length} and the critical Reynolds number $Re_{\theta c}$ are two important relations in this model. Langtry correlates them with the transition Reynolds number $Re_{\theta t}$ in some experimental data, such as (5.23) and (5.24), in which, these two empirical relations are directly correlated to the accuracy of the transitional numerical simulation. The correlation transition model also includes many other parameters in consideration of the transition caused by the separation flow. The corresponding boundary conditions are required for the intermittent factor γ and the momentum thickness Reynolds number Re_θ at the boundary. The intermittent factor is taken 0 at the non-slip wall surface, and the intermittent factor is taken 1 in the free flow. The flux of Re_θ of Momentum Thickness Re_θ is 0 at the wall surface, which can be calculated according to the empirical relationship based on the degree of turbulence in the free flow.

$$Re_{\theta c}=f\left(\overline{Re_{\theta t}}\right) \tag{5.23}$$

$$F_{length}=f\left(\overline{Re_{\theta t}}\right) \tag{5.24}$$

$$P_{k,new}=\gamma_{eff}\cdot P_{k,orig} \tag{5.25}$$

In the equation:

$Re_{\theta c}$—the critical Reynolds number where the intermittent factor starts to increase;

F_{length}—the length of transition area;

γ_{eff}—Effective Intermittent Factor;

$P_{\text{k, orig}}$—generation item of original turbulent kinetic energy equation;

$P_{\text{k, new}}$—Generation Item of New Turbulent Kinetic Energy Equation.

This type of new transition model has been applied in the simulation of passing flow of a considerable number of civil aircraft, which has achieved some gratifying results. It has been continuously adopted in the numerical simulation of supersonic and hypersonic transition flows to provide more detailed aerodynamic data for aircraft. Shang Qing and others from China Academy of Aerospace Aerodynamic Technology conducted a numerical simulation study with the transition model on the transitional turbulent flow of the blunt-head double-wedge hypersonic flow. Furthermore, the numerical gradient of the density gradient numerical schlieren under different flow conditions.

The pressure coefficient distribution and heat flow Stanton number distribution obtained under different flow conditions. The wall temperature is 300 K in the calculation, and the flow state includes laminar flow, transition, and total turbulence. It can be seen from the figure that the size of the laminar flow separation area is larger, the separation shock caused is near the front, the transition flow separation area is reduced, the separation shock moves backward, and the flow separation is almost not seen in the whole turbulent flow. The results that is calculated according to the laminar flow state or the full turbulent state can not correctly reflect the wind tunnel test situation. Only the obtained results from the numerical simulation according to the transition state are in best agreement with the test data, and can accurately simulate the interaction between transition and flow separation, that is, the flow separation occurs in the area near the corner of the double wedge, the flow separation induces a transition, and the transition affects the flow separation.

They also carried out numerical simulation researches on the separation flow and transition of the three-dimensional inlet with the prediction method. There are two cases selected for the incoming turbulence, namely Tu 0.05 and 3.0%, which correspond to the state of static wind tunnel and the state of noise wind tunnel, respectively. The model streamlines that is calculated according to the state of the quiet wind tunnel and the state of the noise wind tunnel. It can be seen from it that there is a three-dimensional flow separation near the corner of the intake port in the state of the quiet wind tunnel. The separation flow area becomes smaller in the state of the noise wind tunnel, and the occurrence of transition weakens the strength of flow separation.

The heat flow distribution on model surface calculated according to the state of the static wind tunnel and the state of the noise wind tunnel. And the surface heat flow on the second compression surface is low in the state of the static wind tunnel and there is no change the transition occurs. The heat flow on the second compression surface is significantly increased in the state of the noise wind tunnel, which indicates that there is a transition phenomenon. The heat flow on the object surface will increase after the transition occurs.

Then, Fu Song et al. established a new k-ω-γ turbulence/transition model based on the Reynolds average method that can reasonably reflect the effects of disturbance modes and compressibility. It consists of three transport equations about intermittent factor γ, pulsating kinetic energy k and its unit dissipation rate ω. The pulsating kinetic energy k includes the turbulent pulsating part and the non-turbulent pulsating part in this model. The modeling of the non-turbulent pulsating adopts the results of stability analysis. The source item of the intermittent factor γ equation is constructed with a function that automatically determines the starting position of transition, in which all expressions are composed of local variables, and can also be easily applied to modern CFD programs. Reduced to the standard SST turbulence model in the complete turbulence region, the model has been validated in the boundary layer flow under subsonic, supersonic and hypersonic conditions, and can be applied to a wider range of Mach number.

$$\frac{\partial(\rho\gamma)}{\partial t} + \frac{\partial(\rho U_j \gamma)}{\partial x_j} = \frac{\partial}{\partial x_j}\left[(\mu + \mu_{eff})\frac{\partial\gamma}{\partial x_j}\right] + P_\gamma - \varepsilon_\gamma \tag{5.26}$$

$$P_\gamma = C_4 \rho F_{onset}\left[-ln(1-\gamma)\right]^{0.5}\left(1 + C_5 \frac{k^{0.5}}{(2E_u)^{0.5}}\right)\frac{d}{v}|\nabla E_u| \tag{5.27}$$

$$\varepsilon_\gamma = \gamma P_\gamma \tag{5.28}$$

In this equation:

ρ—Density (kg/m^3);

γ—Intermittent Factor;

U_j—Speed in j Direction (m/s);

μ—Viscosity Coefficient (kg / (m-s));

μ_{eff}—Effective Viscosity Coefficient (kg/ (m · s));

P_γ—Generation Item of Intermittent Factor Equation;

ε_γ—the Dissipative Item of the Intermittent Factor Equation;

F_{onset}—The Judgment at the Starting of the Transition;

k—Turbulent Flow Energy ((m/s) 2);

E_u—the Average Flow Kinetic Energy of the Local Flow Relative to the Wall ((m/s) 2);

d—Object Range (m);

v—Motion Viscosity Coefficient (m2/s);

x_j—Coordinate in the Direction of j;

t—Time (s);

C_4, C_5—Model Constants.

Figure 5.10 indicates the distribution of temperature on the surface of the sharp cone model and the comparison with the test and the calculation results of the Hassan model. The test data is from the half cone angle 5 in the low turbulence wind tunnel of NASA Langley Research Center.

The incoming Ma is 3.5 in the cone test. Figure 5.11 indicates the distribution of the surface temperature of the conical model with a half-cone angle of 5° and a skirt tail conducted in the low-turbulence wind tunnel of NASA Langley Research Center. Figure 5.12 indicates the comparison between the calculated wall law of the conical boundary layer and the theoretical value.

The transition process is affected by many factors, in which there are multiple transition mechanisms. The current transition models have not been able to integrate multiple mechanisms into one model framework. Most of the transition models adopt the Reynolds average method, which cannot fully reflect the growth process of unstable disturbance waves. And the simulation calculation of the transition flow development process will be more or less biased. The transition model should be more suitable for parallel calculation of complex-shaped objects under the condition that the current large-scale calculation conditions are relatively mature, and the calculation accuracy and the convergence of the solution must be guaranteed. It is acceptable in engineering applications although the current transition model is not accurate enough to simulate the length of the transition region. The transition model needs to be continuously improved for in-depth structural research conducted on the transition mechanism and the scope of application continuously expanded, so as to obtain greater application prospects in the aerospace field.

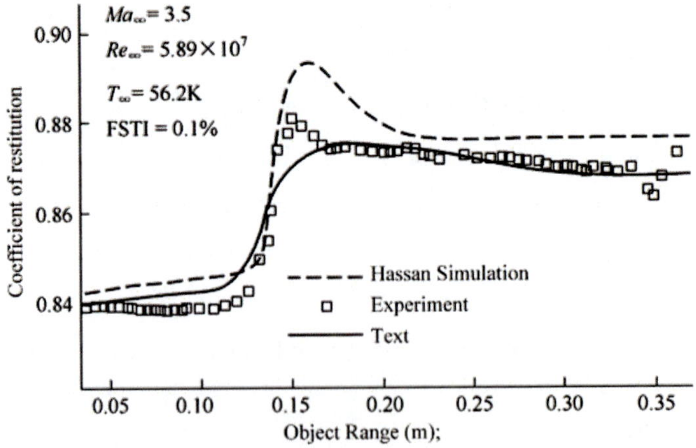

Fig. 5.10 Temperature recovery coefficient of a cone with zero angle of attack [39]

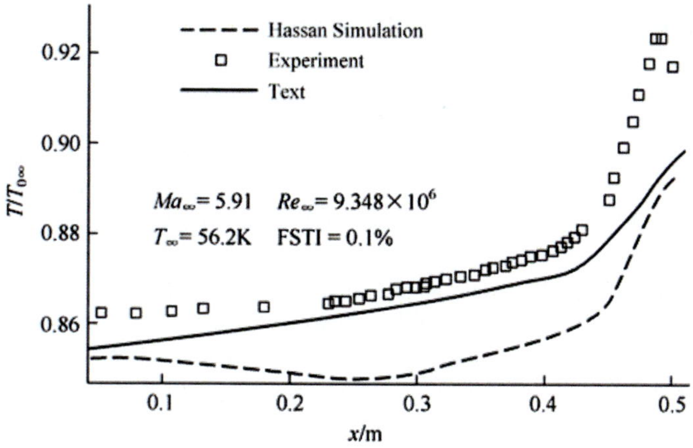

Fig. 5.11 Temperature distribution of hypersonic cone surface [39]

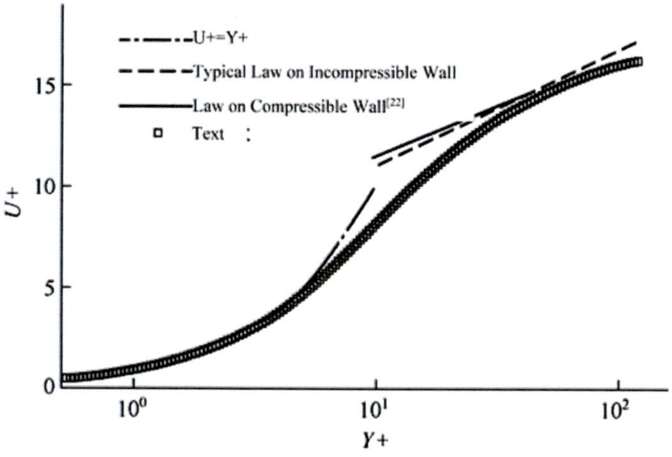

Fig. 5.12 The inner velocity distribution of fully developed turbulent section of hypersonic cone [39]

5.5 Engineering Prediction Method Based on Transition Criteria

Transition Criterion refers to a simple algebraic relationship established to determine the location of transition. A lot of relevant work has been carried out in previous studies on various types of aircraft with outstanding transition turbulence problems, such as flat plate problems, cone problems, blunt-shaped reentry vehicles, space shuttles, etc., and some transition criteria have been established. It has helped many engineering designers to carry out the design work of the aircraft since the transition criterion is simple and practical. However, each transition criterion has a certain scope of application, and there is no universal transition criterion as transition is a complex and non-linear problem with many influencing factors.

The transition criteria at early stage were mostly developed and verified based on the data of the low-speed wind tunnel test, which is basically reliable for the natural transition of the flow area before transonic speed, but it is not applicable for high-speed flow. However, there is a small number of transition criteria for the development of high-speed flow often based on conventional wind tunnel test data and very few flight test data, which do not consider the effects of temperature, noise, and pressure gradient. Therefore, these transition criteria cannot be adopted to predict natural transitions on the surface of hypersonic vehicles. The mechanism is more complex and the theory is more incomplete for manual transitions. It is necessary to carry out experiments well designed and obtain a large number of experimental data, so as to establish practical transition criteria to predict the transition of similar problems.

The transition method is basically similar at present. A large number of ground tests and flight tests are carried out to obtain data after the aircraft shape and flight corridor are determined. The empirical relationship for judging the transition can be obtained after analysis, summary, and refining. These transition criteria are adopted to predict the transition situation of the boundary layer of other similar aircraft without test data. Therefore, the transition rules established so far generally are only applicable to aircraft of similar contours. Here it introduces some transition criteria of hypersonic boundary layer and its data sources, application scope, which is expected to help engineering designers.

The transition criterion of Potter & Whitfield was proposed by Potter and Whitfield in 1962 [41]. It can be applied in the contour of a pointed cone. It is in good agreement compared with the results of the Ma 7 for pointed cone wind tunnel test conducted by ONERA, but there are differences between the results for a particularly small height rough element. The criterion achieves the prediction of the transition by correlating the roughness Reynolds number with the transition position. A prediction method is proposed for evaluating the transition position change caused by roughness based on this. The modified Reynolds number is:

$$Re_{kk} = \frac{\rho_k k U_k}{\mu_k}\left(\frac{T_k}{T_w}\right)^{\omega+0.5} \tag{5.29}$$

In this equation:

Re_{kk}—Rough Element Height Reynolds Number;

ρ_k—Density (lb/inch^3);

kRough Element Height (inch);

U_k—Speed (inch/s);

μ_k—Viscosity Coefficient (lb/ (inch · s));

T_k—Temperature at the position of rough element (K);

T_w—Wall Temperature (K);

ωConstant, 0.76.

The transition criterion of Van Driest & Blumer was proposed by Van Driest and Blumer in 1967 [42], which was based on test data with isolated spherical rough elements distributed on the cone surface. The transition criterion is not in good agreement with the cone test data. NASA revised this criterion in 1997 and adopted it in the transition prediction of the space shuttle flight test and wind tunnel test.The degree is still not too good of agreement between the prediction result and the test data, of which the error of the wind tunnel test data is ±65%, and the errors of flight test data are ±69% (STS-28) and ±208% (STS-73).

$$Re_{\delta_1} = 242\left(1 + \frac{\gamma - 1}{2}M_e^2\right)\left(\frac{k}{\delta_1}\frac{1}{1 + B/B_s}\right)^{-1.1} \tag{5.30}$$

In this equation:

$Re_{\delta 1}$—Reynolds Number of Boundary Layer Thickness;

γ—Gas Constant;

M_e—Mach Number at the Outer Edge of the Boundary Layer;

B—Flow Velocity Gradient (s^{-1}) at the Outer Edge of the Boundary Layer.

B_S—Speed Gradient at the Stagnation Point (s^{-1}), which is taken as 0.5;

k—Rough Element Height (inch);

δ_1—Boundary Layer Thickness (inch).

The transition criterion of PANT was proposed by Wool in 1975 [43], which was based on the results of the supersonic and hypersonic wind tunnel tests in the "Passive Nosetip Technology" project. This criterion can be adopted to predict the transition of the shape of the blunt body. The predicted transition position always exited in the subsonic area, so it is not suitable for the contour of a sharp cone or a small blunt cone. Then it is necessary to combine other conditions for other contours of aircrafts, such as redefinition of the Reynolds number considering the radius of the head or the corresponding geometric size.

$$\psi_1 = Re_\theta\left(\frac{kT_e}{\theta T_w}\right)^{0.7} 1 < \frac{kT_e}{\theta T_w} < 10, \tag{5.31}$$

In the equation:

ψ_1—Transition Criterion, the transition takes place when the value is 215;

Re_θ—Reynolds Number of Momentum Thickness of Boundary Layer;

k—The Peak-valley Difference of Rough Element (mil);

T_e—the Temperature of the Outer Edge of the Boundary Layer °R);

T_w—Wall Temperature °R);

θ—Boundary Layer Momentum Thickness (mil).

The transition criterion of Boudreau was proposed by Boudreau in 1981 [44] and can be adopted to predict the transition of the cone surface caused by the roughness of the head.

$$\lambda_M^* = X\frac{k}{r_n}\left(\frac{P_E}{P_o}\right)^{0.9}\left(Re_{\infty,r_n}\right)^{1.2} \tag{5.32}$$

$$X = Re_{e,\theta}^*\left[\frac{T_e^* r_n}{T_w \theta^*}\right]^{0.7}\left(Re_{\infty,r_n}\right)^{-0.85} \tag{5.33}$$

In this equation:

$\lambda^*{}_M$—Transition Criterion, $\lambda * {}_{\mathrm{M}} \geq 800$, the Transition Takes Place;

r_n—The radius of the head of the front edge of the model (inch);

θ^*—Boundary Layer Momentum Thickness at the Speed of Sound (inch);

$Re_{\infty,rn}$—Reynolds number of Free Flow based on the Radius of the Head of the Leading Edge;

$Re^*{}_{e,\theta}$—Momentum Thickness Reynolds Number at the Outer Edge of the Boundary Layer at the Speed of Sound;

k—Rough Height (inch);

P_E—Rough Pressure (psia);

P_0—Pitot Pressure of Free Flow (psia);

T^*—The temperature of the outer edge of the boundary layer at the speed of sound (°R);

T_w—Wall Temperature °R).

The transition criterion of NASA was put forward by Bertin et al. in 1981 [45]. The criterion is specifically for the space shuttle type appearance. It has been indicated in some researches that the relevant parameters must be adjusted for other geometric contours. Bouslog et al. [46] proposed a new transition criterion in 1997 which combined the specific conditions of the space shuttle and the PANT criterion and is more suitable for predicting the transition situation of the space shuttle. Then Berry et al. [47] simplified it and put forward a simple version of the transition criterion in 1998.

$$\frac{Re_\theta}{M_e} = 191\left[\frac{k}{\theta}\frac{T_e}{T_w}\frac{1}{1 + B/B_s}\right]^{-0.444} \tag{5.34}$$

In this equation:

Re_θ—Reynolds Number of Momentum Thickness of Boundary Layer;

M_e—Mach Number at the Outer Edge of the Boundary Layer;

B—Flow Velocity Gradient (s^{-1}) at the Outer Edge of the Boundary Layer.

B_s——Speed Gradient at the Stagnation Point (s^{-1}), which is taken as 0.5;

k—Rough Element Height (inch);

θ—Boundary Layer Momentum Thickness (mil);

T_e—the Temperature of the Outer Edge of the Boundary Layer °R);

T_w—Wall Temperature °R).

The transition criterion of BATT & LEGNER [48] is based on the test data and smooth wall transition data in the PANT project, which attempts to simulate natural transition and artificial transition at the same time. The criterion took two cases in consideration, the first case corresponds to a smooth wall transition, and the second case corresponds to a rough wall transition.

However, this criterion fails to fully consider the pressure gradient and temperature effects, therefore it is not recommended for predicting natural transitions.

$$Re_{\theta T} = \begin{cases} 500X^{-1.5}, & 1 \le X \le 10 \\ 500, & X < 1 \end{cases} \tag{5.35}$$

$$X = \frac{kT_e}{\theta T_w}\frac{1}{1 + 350k/R_n} \tag{5.36}$$

In the equation:

$Re_{\theta T}$—Reynolds Number of Transitional Momentum Thickness;

R_n—The radius of the head of the front edge of the model (inch);

k—Rough Element Height (inch);

θ—Boundary Layer Momentum Thickness (inch).

T_e—the Temperature of the Outer Edge of the Boundary Layer (°R);

T_w—Wall Temperature (°R).

The transition criterion of BLT [49] was proposed in the NASP project in the United States, which is mainly based on flight test data, partly theoretical prediction data and static wind tunnel test data. The criterion consists of BLT-1A, BLT-1B and BLT-1C. The criterion of BLT-1A is applicable to the contour of a sharp cone, which was later improved to take the transition of blunt cone and flat boundary layer into consideration in the form of a display factor to form the BLT-1B criterion, which is suitable for the prediction of slender body transition.

The criterion of BLT-1C are developed for NASP aircraft, and also consider various influencing factors in the form of display factors, which are suitable for the prediction of the transition of the fuselage, air intake and tail nozzle.

The form of BLT-1A criterion is:

$$\left(\frac{Re_\theta}{M_e}\right)_{sharp} = 318 \tag{5.37}$$

In this equation:

Re_θ—Reynolds Number of Momentum Thickness;

M_e—Mach Number at the Outer Edge of the Boundary Layer;

The correction form is:

$$\left(\frac{Re_\theta}{M_e}\right)_{effective} = \left(\frac{Re_\theta}{M_e}\right)_{sharp} \cdot A \cdot B \cdot P \cdot T \cdot G \cdots \tag{5.38}$$

In this equation:

A—Wedge Angle Factor;

B—Bluntness Factor;

P—Pressure Gradient Factor;

T—Wall Temperature Factor;

G—Three-dimensional Effect Factor.

Other transition influencing factors can also be taken into consideration in the form of display factors. Among them, B has the following form for the contour of the blunt body:

$$B = \frac{(Re_\theta/M_e)_{blunt}}{(Re_\theta/M_e)_{sharp}} = K_1 + K_2\left[(Re_{rn})_\infty\right]^{K_3} \tag{5.39}$$

In the equation:

Re_{rn}—Reynolds Number Based on the Radius of the Head of the Leading Edge;

∞—Indicating the State of Free Flow;

K_1, K_2, K_3—Constants (Table 5.1).

Reda et al. [50] proposed the transition criterion caused by roughness of the ablative bluff body heat shield during atmospheric reentry in 2007, in which three conditions were taken into consideration, including boundary layer transition on smooth wall, boundary layer transition within the critical roughness range, and boundary layer transition under the asymptotic condition of high roughness/low Reynolds

Table 5.1 Table of constant in BLT-1C transition criterion [49]

		K_1	K_2	K_3
$M_\infty \leq 6$		1.0	0. 0854	0.32
$6 < M_\infty < 15$		1.0	0. 1082	0.25
$M_\infty \geq 15$	$Re_{\text{m}} \leq 5000$	1.0	-1.9794×10^{-5}	1.0
	$Re_{\text{mn}} > 5000$	0. 8511	9.986×10^{-6}	1.0

number. The form from Reda is $Re_k = \rho_k u_k \overline{k}/\mu_w$. Schneider also provided a form in 2007 [51], that is $Re_{kk} = \rho_k u_k \overline{k}/\mu_k$. Reda believes that at this time $\mu_k \approx \mu_w$, the two forms of Re_k should be consistent for small roughness and approximate adiabatic conditions. The viscosity changes greatly and the viscosity change in the boundary layer reaches 45% in the HIFiRE project, since the temperature of the frustum is only 15% of the total temperature, so the two forms of Rek give different results. SchneideF [51] also analyzed the data from Braslow. It is indicated in the research that $\sqrt{Re_k}$ changes in the range of 15–45 in the range of Mach 4 or less, and $\sqrt{Re_k}$ suddenly jumps to 200 in the range of Mach 4–6.

References

1. Van Ingen JL (1956) A suggested semi-empirical method for the calculation of the boundary layer transition region. University of Technology, Department of Aerospace Engineering, Report UTH-74
2. Van Ingen JL (1965) Theoretical and experimental investigations of incompressible laminar boundary layers with and without suction. University of Technology, Department of Aerospace Engineering, Report UTH-74
3. Van Ingen JL (1976) Transition, pressure gradient, suction, separation and stability theory. Low-speed boundary-layer transition workshop II
4. Johnson HB, Candler GV (2006) Analysis of laminar-turbulent transition in hypersonic flight using PSE- Chem. In: AIAA 36th (Fluid Dynamics Conference and Exhibit), vol 5. San Francisco, CA, pp 5–8
5. Mack LM (1977) Transition and laminar instability. Jet Propulsion Lab 5:77–15
6. Dryden HL (1959) Transition from laminar to turbulent flow. In: Lin CC (ed) Turbulent flows and heat transfer. Princeton University Press, New York
7. Jisheng L (2015) Transition and prediction of hypersonic boundary layer. J Aeronaut 36(1):357–372
8. Smagorinsky J (1963) General circulation experiments with primitive equations. I. Basic Experiment Mon, Weather Rev 91:99–164
9. Chasnov JR (1991) Simulation of the Kolmogorov inertial subrange using an improved subgird model. Phys Fluids A 3:188–200
10. Deardorff JW (1970) A numerical study of three-dimensional turbulent channel flow at large Reynolds numbers. J Fluid Mech 41:453–480
11. Schumann U (1975) Subgrid scale model for finite difference simulations of turbulent flow in plane channels and annuli. Comp Phys 18:376–404
12. Clark RA, Ferziger JH, Reynolds WC (1977) Evaluation of subgrid-scale turbulence models using a fully simulated turbulent flow. NASA-CR-152642
13. Moin P, Kim J (1982) Numerical investigation of turbulent channel flow. Fluid Mech 12(5):118

14. Bardina J, Lombard CK (1985) Three dimensional CSCM method for the compressible navier–stokes equations with application to a multi-nozzle exhaust flowfield. AIAA-85-1193
15. Germano M, Piomelli U, Moin P et al (1991) A dynamic subgrid-scale eddy viscosity model. Phys Fluids A 3:1760–1765
16. Zang TA, Chang C Transition prediction toolkit: LST, SIT, PSE, DNS, and LES. N93-27429/8/XAD
17. Vreman AW, Sandham ND, Luo KH (1996) Compressible mixing layer growth rate and turbulence characteristics. J Fluid Mech 320:235–258
18. Zhibin Z, Xiangjiang Y, Lin C (2015) Partition parallel algorithm of high-order compact format. J Comput Mech 32(6):825–830
19. Fang J, Lu L, Shao L (2009) Large Eddy simulation of turbulent flow in compressible channels with wall vibration. Sci China (G Ser: Phys, Astron) 39(6):843–853
20. Baldwin B, Lomax H (1978) Thin layer approximation and algebraic model for separated turbulent flows. AIAA Paper 257
21. McDonald H, Fish R (1973) Practical calculations of transitional boundary layers. Int J Heat Mass Transf 16:1729–1744
22. Arad E, Berger M, Israeli M et al (1982) Numerical calculation of transitional boundary layers. Int J Numer Methods Fluids 2:1–23
23. Priddin CH (1974) The behavior of the turbulent boundary layer on curved porous walls. PhD thesis: Imperial College
24. Scheuerer G (1983) Entwicking eines verfahrens zur berechnung zweidimensionaler, grenzschichten an gastur- binensschaufeln. University Karlsruhe, Karlsruhe
25. Craft TJ, Launder BE, Suga K (1997) Prediction of turbulent transitional phenomena with a nonlinear eddy-viscosity model. Int J Heat Fluid Flow 18:15–28
26. Hadzic I (1991) Second-moment closure modeling of transition and unsteady turbulent flows. PhD thesis: University of Delft
27. Wilcox D (1992) Turbulence Modeling for CFD[D]. CA, DCW Industries, La Canada
28. Schmidt R (1987) Two-equation low-Reynolds-number turbulence modeling of transitional boundary layer flow characteristic of gas turbine blades. PhD thesis: University of Minesota
29. Johnson K (1987) Heat transfer calculations on turbine airfoils. Master's thesis: The University of Texas at Austin
30. Schmidt R, Patankar S (1991) Simulation boundary layer transition with low-Reynolds-number k-e turbulence models: part 2-an approach to improving the predictions. J Turbomach 113:13–28
31. Libby PA (1975) On the prediction of intermittent turbulent flows. J Fluids Mech 68:273–295
32. Dopazo C (1977) On conditioned averages for intermittent turbulent flows. J Fluids Mech 81:433–438
33. Steelant J, Dick E (2001) Modeling of laminar-turbulent transition for high freestream turbulence. J Fluids Eng 123:22–30
34. Huang PG, Suzen YB (2000) An intermittency transport equation for modeling flow transition. AIAA Paper 2000-0287
35. Bradshaw P (1994) Turbulence: the chief outstanding difficulty of our subject. Exp Fluids 16:203–216
36. Walters DK, Leylek JH (2000) Impact of film—cooling jets on turbine aerodynamic lossed. ASME J Turbomach 122:537–545
37. Volino RJ (1998) A new model for free-stream turbulence effects on boundary layers. ASME J Turbomach 120:613–620
38. Menter FR, Langtry RB, Likki SR, et al (2004) A correlation based transition model using local variables part 1—model formulation. ASME-GT2004-53453, ASME TURBO expo 2004, Vienna, Austria
39. Fu S, Wang L (2009) Simulation of the transition of hypersonic boundary layer based on reynolds average method. Sci China (G Ser: Phys Mech Astron), 39(4):617–626
40. Langtry RB, Menter FR (2005) Transition modeling for general CFD applications in aeronautics. AIAA 2005-0522

41. Potter J, Jack D (1962) Effects of slight nose bluntness and roughness on boundary-layer transition in supersonic flows. J Fluid Mech 12(4):501–535
42. van Driest ER, Blumer CB (1963) Boundary layer transition: freestream turbulence and pressure gradient effects. AIAA J 1(6):1303–1306
43. Wool MR (1975) Passive nosetip technology (PANT) program. ADA019186
44. Boudreau AH (1981) Correlation of artificially induced boundary-layer transition data at hypersonic speeds. J Spacecr Rocket 18(2):152–156
45. Bertin JJ, Hayder TE, Goodrich WD (1981) Comparison of correlations of shuttle boundary-layer transition due to distributed roughness. AIAA 19th (Aerospace Sciences Meeting), St. Louis, Missouri 13(1):12–15
46. Bouslog SA, Bertin JJ, Berry SA, et al (1997) Isolated roughness induced boundary-layer transition: shuttle orbiter ground tests and flight experience. AIAA, Aerosp Sci Meet & Exhib, 35th, Reno, NV, 22(1):6–9
47. Berry SA, Bouslog SA, Brauckmann GJ et al (1998) Shuttle orbiter experimental boundary-layer transition results with isolated roughness. J Spacecr Rocket 35(3):241–248
48. Batt RG, Legner HH (1983) A review of roughness-induced nosetip transition. AIAA J 21(1):55–57
49. Lau KY (2008) Hypersonic boundary-layer transition: application to high speed vehicle design. J Spacecr Rocket 45(2):176–183
50. Reda DC, Wilder MC, Bogdanoff DW, et al (2007) Transition experiments on blunt bodies with distributed roughness in hypersonic free flight. AIAA 45th (Aerospace Sciences Meeting and Exhibit), Jan 8–11, Reno, Nevada, 2(3): 306
51. Schneider SP (2007) Effects of roughness on hypersonic boundary-layer transition. AIAA 45th (Aerospace Sciences Meeting and Exhibit), Jan 8–11, Reno, Nevada, 21(3):305

Chapter 6
Embedded Aerosphere Sensing System

6.1 An Overview

As a comprehensive measurement system, the atmospheric data system processes and calculates other parameters related to atmospheric data, such as barometric altitude, height deviation, lifting speed, true airspeed, indicating airspeed, Mach number, rate of change of Mach number, atmospheric density ratio, etc., based on the basic raw information measured by the sensors, such as static pressure, dynamic pressure, angle of attack, and total temperature, which will be sent to flight control system, navigation system, engine control system and flight data recording system that is one of the necessary avionics subsystems for various aircraft. The system generally consists of three parts: the sensor part that provides the original information, the task computer or compensation solver, and the signal output device. The basis of the atmospheric data system is the sensor, which is sensitive to the information of the air flow, such as the total pressure, static pressure, total atmospheric temperature, angle of attack and angle of sideslip. The mission computer carries out comprehensive calculation processing according to the correlation equation between the parameters of air pressure height, lifting speed, corrected airspeed, true airspeed, Mach number and total pressure, total atmospheric temperature etc., so as to obtain the above atmospheric data parameters [1–9].

6.1.1 Traditional Atmospheric Data System

The traditional atmospheric data sensing system takes the airspeed tube as basis, plus a total temperature sensor, angle of attack sensor and side slip angle sensor to measure five parameters of total pressure, static pressure, total atmospheric temperature, angle of attack and side slip angle and then calculate other atmospheric data based on these five parameters. The airspeed head is usually installed at the front of the aircraft and

M. Zhao, *Prediction and Validation Technologies of Aerodynamic Force and Heat for Hypersonic Vehicle Design*, Springer Aerospace Technology,
https://doi.org/10.1007/978-981-33-6526-1_6

extends outward, and the angle of attack sensor is also installed outside the surface of the aircraft head. The impact of the angle of attack on the calibration of the static pressure measurement is difficult to solve when flying at a large angle of attack, using this atmospheric data system, and the measurement accuracy of the angle of attack, etc. is relatively low. The measurement accuracy will decrease significantly with the increase of the angle of attack, and the measurement of sideslip Angle is avoided in many flight control systems. This avoidance inevitably leads to a reduction in aircraft performance. The airflow will show the characteristics of hypersonic airflow when the aircraft is flying in hypersonic speed. The shock layer is thin, the boundary layer becomes thick, and viscous interaction occurs. Part of the kinetic energy of the airflow is converted into gas internal energy when the hypersonic airflow decelerates by shock wave compression or viscous block, that is, the temperature of the airflow increases, and components such as airspeed head and angle of attack sensors outside the surface of the aircraft cannot withstand such a high heat environment. Stealth aircraft require to avoid "bi-facial unit" as much as possible, that is, two metal surfaces that are perpendicular to each other, since it will cause the radar signal to be "secondarily reflected" and return in the direction opposite to the incident direction, which is extremely harmful to stealth. And similar "angular reflector" or "concave structure" will cause multiple reflections of the radar signal. It is a better way to make the lower surface of the aircraft as flat as possible. The surface of the aircraft adopts a surface with a uniform change in the radius of curvature. The airspeed head, angle of attack sensor, and slip angle sensor will form some vertical and concave surfaces to the surface of the aircraft, which is hardly to meet the needs of aircraft stealth. It can be seen obviously that the traditional atmospheric data system has failed to meet the needs from modern advanced aircraft, and it is necessary to research and develop a new atmospheric data system [8].

The part of sensor of the traditional atmospheric data sensor system that is sensitive to the original information usually consists of the airspeed head (static pressure and total pressure sensor), the total atmospheric temperature sensor, the angle of attack sensor and the side slip angle sensor.

The computer carries out resolution and calculation on various atmospheric data parameters, and finally outputs them to various required avionic systems in a certain form. The basic composition is shown in Fig. 6.1 [8].

6.1.1.1 Airspeed Head

The airspeed head, also known as pitot tube, was invented by French H. Pitot, which is a tubular device that measures the total pressure and static pressure of airflow. The structure of the airspeed head is shown in Fig. 6.2. The head is hemispherical, the air flow is transmitted by a double-layer casing, and the back is connected to the temperature sensor. The head is in the direction of the incoming flow during measurement. The airflow is introduced through the small hole (total pressure hole) at the center of the head and transmitted to the temperature sensor (total pressure sensor) through the inner tube. The total pressure of the inflow could be obtained

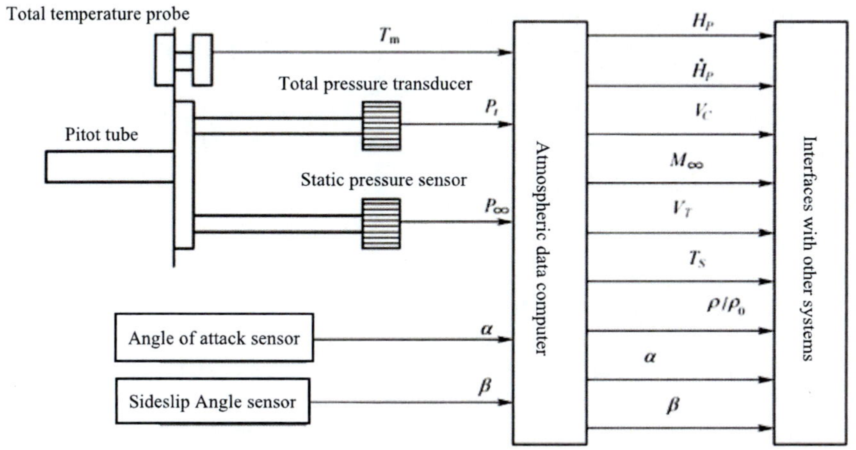

Fig. 6.1 Basic composition of traditional atmospheric data system

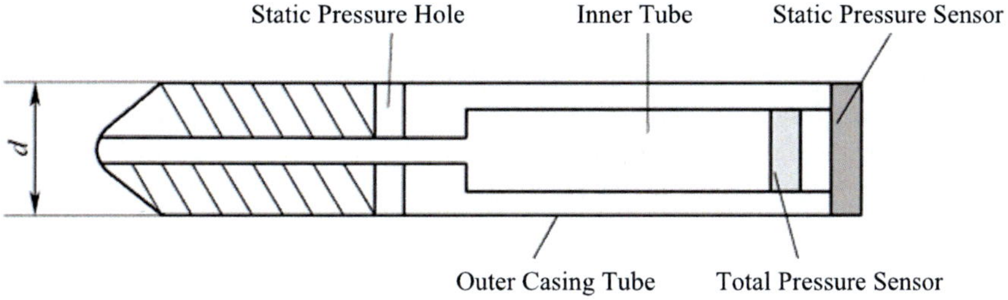

Fig. 6.2 Schematic diagram of airspeed head structure

from sensitivity. There is a row of static pressure holes, and the air flow is transmitted to the temperature sensor (static pressure sensor) through the outer casing, and the incoming static pressure P_∞ is obtained after sensitivity [8].

A vibration pressure sensor or a solid membrane pressure sensor is usually adopted as the pressure sensor that is used to measure the static pressure. The pressure sensor of vibration type measures the pressure according to the principle that its natural frequency will change when the pressure received by the mechanical oscillation system changes. The principle of sensitivity on pressure of the solid membrane capsule pressure sensor is that the vibration membrane in the membrane capsule occurs under the pressure bending.

The position for installation of the airspeed head should be selected in the area outside the aircraft where the airflow is less affected by the aircraft. Wind tunnel calibration and flight calibration on the aircraft are required after the airspeed head is installed. The static pressure and the magnitude of the total pressure error varies with the Mach number and angle of attack due to the different aircraft and installation location. Therefore, it cannot be interchanged arbitrarily in different aircraft or different positions of the same aircraft after the airspeed head is installed and calibrated. It is generally installed directly in front of the head of the aircraft, such as the Russian Su-27 and the Chinese J-1. Same with American F-22, it can also be installed in other locations, for example, the Thunder jointly developed by China and Pakistan is installed on the left side of the nose, the Alpha jet trainer jointly developed by Germany and France is also installed on the left, and the Chinese K8 trainer is installed on the upper side of the nose.

6.1.1.2 Total Temperature Sensor

Objects will block the airflow in the high-speed airflow, so that the kinetic energy of the airflow is converted into heat energy, which results in the local temperature of the airflow to increase. The components of the aircraft exposed to the high-speed airflow will cause the temperature rises of the airflow due to this blocking effect on the airflow. The elevated temperature needs to be estimated to ensure these components in a normal operating range. What's more, the calculation on some atmospheric data also requires the static temperature of the airflow, and the static temperature is calculated by the total temperature. Therefore, the total temperature of the air flow flowing in high speed needs to be measured. The element used to measure the fully blocked temperature of the high-speed airflow (that is the total temperature) is called the total temperature sensor [8].

The schematic diagram for the total temperature sensor structure is shown in Fig. 6.3. It mainly consists of a duct, a boss, a sensitive element and a pillar. The duct is used to introduce airflow. The boss is located at the smallest cross-sectional

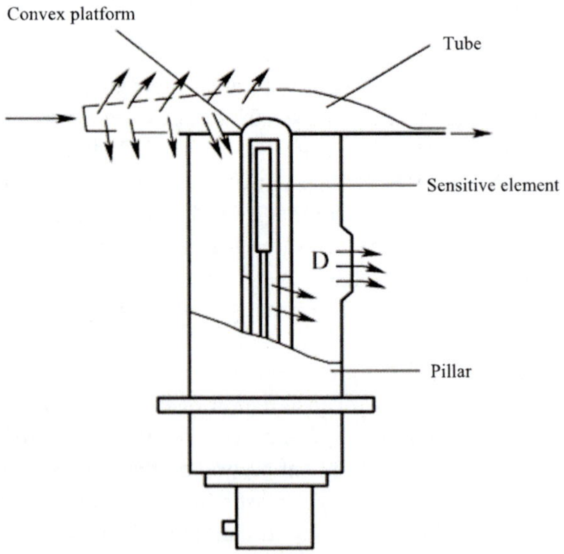

Fig. 6.3 Structure diagram of total temperature sensor

area of the duct. A boundary layer will be formed on the inner wall of the tube when the airflow flows through here due to the viscosity of the gas. The airflow is blocked, and the airflow speed is quickly reduced to a minimum. The blocking process of the airflow is too short to exchange energy with the outside world. The process is adiabatic, and the kinetic energy of the airflow is all converted into heat energy, which causes the temperature of the inner wall of the tube to rise. The arrows in the figure indicate the direction of the airflow. The airflow is blocked by the boss and enters the temperature sensing element compartment. The sensitive element is sensitive to the blocked temperature of the airflow, which is the total temperature of the airflow. The sensitive element is usually heat resistance made of a platinum wire (also used Nickel wire) [8].

6.1.1.3 Sensors of Angle of Attack and Sideslip Angle

The angle of attack is the angle between the projection of the velocity vector on the aircraft symmetry plane and the body axis. It directly affects the lift and drag of the aircraft as an essential flight parameter. It may result in insufficient lift of the aircraft if the angle of attack is not properly controlled. A considerable number of flight accidents occur during the take-off and landing phases, many of which are caused by stalls caused by the angle of attack out of control. What's more, the angle of attack also affects the total pressure and accuracy of static pressure measurement. The device used to sense, convert and output the angle of attack information is called the sensor of angle of attack. The angle of attack sensor can be roughly divided into three types according to different working principles: rotating wind vane type sensor of angle of attack, differential pressure tube type sensor of angle of attack, and zero differential pressure sensor of angle of attack [9].

The principle of rotating wind vane sensor of angle of attack is shown in Fig. 6.4. It mainly consists of blades, angle converters and rotating shafts. The blade acts as a weather vane in the measurement and is used to sense the aerodynamic effect to measure the angle of attack of the aircraft. The rotating shaft mainly plays a

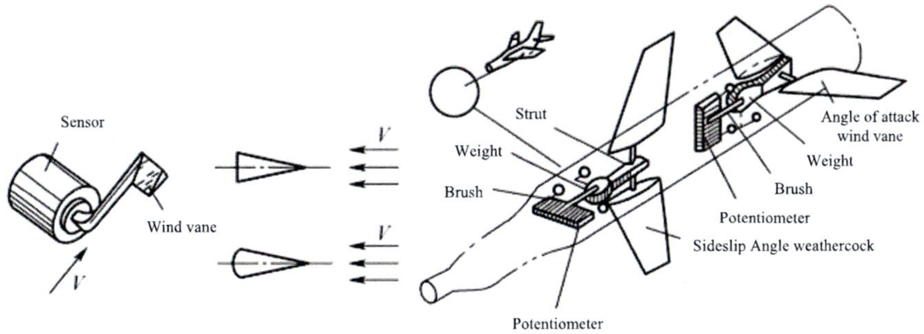

Fig. 6.4 Rotating wind vane sensor of angle of attack

supporting role and rotates with the blade. The angle changer converts the rotating angle of the rotating shaft into an electrical signal. The aerodynamic forces on the upper and lower surfaces of the blades are unequal when the axis of the aircraft is inconsistent with the direction of the airflow, so they rotate relative to the axis of the aircraft until the centerline of the blade is consistent with the direction of the airflow. The angle between the line and the axis of the aircraft is the angle of attack of the aircraft, that is the rotation angle of the blade, which the potentiometer converts into electrical signal output.

The rotating wind vane sensor of angle of attack is simple in structure, small in size, and has no principle error, but there is greater influence on the installation position error. However, it is difficult for high-speed aircraft to find a smooth airflow, and the wind vane is easily affected by small disturbances, which is inherently unstable and will cause unstable swing of the weather vane [8].

The differential pressure tube type sensor of angle of attack consists of a differential pressure tube and a pressure sensor. A small hole is opened above and below the axis of the differential pressure tube. The pressure introduced by each hole is equal when the axis of the differential pressure tube is consistent with the direction of the air flow. The pressure is unequal when there is an angle of attack. The Angle of attack can be obtained according to the relationship between the Angle of attack and the pressure difference. It can measure the angle of attack and the angle of sideslip through re-opening holes on the left and right sides of the axis of the differential pressure tube [8].

The zero-pressure differential sensor of angle of attack is a further development of the differential pressure angle-of-attack sensor. As it is shown in Fig. 6.5, it consists of a probe, a gas chamber, a blade, and an angle converter. The probe, angle converter, and blade are all fixed on the hollow shaft. The probe is conical. There are partitions in the middle, and two rows of inlet holes are symmetrically opened on both sides of the centerline. The axis of the probe is parallel to the horizontal axis of the aircraft in flight. The plane of the baffle plate is parallel to the direction of airflow when the angle of attack is zero. The angle is zero between the plane of symmetry of

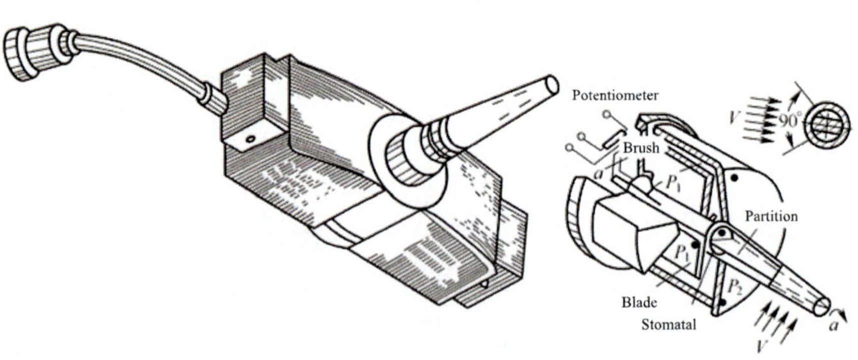

Fig. 6.5 Zero differential pressure sensor of angle of attack

the upper and lower rows of pressure holes and the direction of airflow. The total aerodynamic torque is zero, the hollow shaft does not rotate, and the output signal of the potentiometer is zero. The symmetry planes of the two rows of pressure measuring holes on the probe are inconsistent with the direction of the airflow when the angle of attack is not zero, and the intake groove also rotates relative to the direction of the airflow from the same angle.

The aerodynamic pressure entering the air holes is no longer equal through the two rows of air inlets. The blade is subjected to an aerodynamic moment and drives the hollow shaft and the probe to rotate together. The rotation is not stopped until the symmetry planes of the upper and lower rows of pressure measuring holes are consistent with the direction of the air flow. The angle passed is the angle of attack of the aircraft, and the potentiometer converts it into an electrical signal and outputs it. The zero-differential pressure sensor of angle of attack is a feedback measurement system, so the error is small. The main error comes from the installation position and various friction torques [8].

The principle of the slip angle sensor is almost the same as that of the sensor of angle of attack, except that the installation axis of the sensitive element is perpendicular to the angle of attack sensor.

The installation position has a great influence on the error no matter which angle of attack sensor it is, the location should be found with efforts where the airflow disturbance is small, or use two sensors and install them as much as possible in the plane of symmetry of the aircraft. Generally, the angle of attack side slip angle sensor can be installed on the airspeed tube, so that the air flow is relatively stable, and the angle of attack, side slip angle, total pressure and static pressure can be measured simultaneously, as shown in Fig. 6.6. Otherwise the airspeed probe and static pressure hole can be installed on the wind vane, through which it can measure the angle of attack, side slip angle, total pressure and static pressure at the same time. This

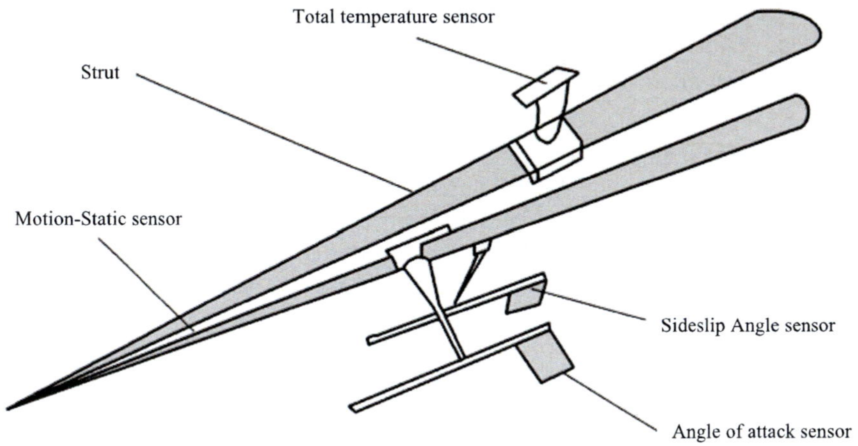

Fig. 6.6 Airspeed tube angle of attack sensor system

integrated airspeed tube/angle of attack beacon system is compact and can be used to greatly reduce the impact of small high angle of attack on pressure measurement. Therefore, it is used in high-performance aircraft, such as the European "Typhoon" fighter, equipped with four sets of such integrated atmospheric data sensor devices. In addition, the side slip angle sensor of angle of attack is affected by the turbulence of the fuselage. It usually measures the local angle of attack at the installation position, not the angle of attack of the free flow in front. Therefore, the true angle of attack of the free flow needs to be obtained by conversion [8].

6.1.1.4 Atmospheric Data Computer

The atmospheric data computer can be used to correct the data measured by the sensor, and calculate other atmospheric parameters, according to the basic atmospheric data sensitive to the sensor, such as total pressure, static pressure, total temperature, and angle of attack, and other atmospheric parameters such as pressure height, lifting speed, indicated airspeed, Mach number, true airspeed, etc. It is a centralized information processing system developed on the basis of simple pneumatic diaphragm instrument and electronic and sensor technology with the advancement of aviation technology. There have been electromechanical analog and electronic meter analog atmospheric data computers in the course of this development, as well as simplified systems with individual single functions, such as altitude speed system, altitude system, angle of attack warning system, etc. Small airborne digital computers and microprocessors were widely used in various airborne equipment in the late 1970s. Atmospheric data computers are usually digital currently in use.

The box of basic principle is shown in Fig. 6.7. It realizes the digital operation of various atmospheric data with high speed and high precision through software [8].

6.1.2 Development Trend of Embedded Atmospheric Data Sensing System

The embedded air data sensing system (Flush Air Data System, FADS) measures the pressure distribution on the surface of the aircraft with rely on the pressure sensor array, and indirectly obtains atmospheric data such as dynamic pressure, total pressure, angle of attack, and sideslip angle through a specific algorithm. FADS has been successfully applied to various types of aircraft from the United States such as X-15, F/A-18, X-31, X-33, X-34, X-38, X-43A, space shuttle and "Mars Science Laboratory" Mars probe, etc. [10–36]. The development of FADS in the United States is shown in Fig. 6.8. Other aviation powers in the world are also conducting FADS technology research in various aircraft.

The National Aeronautics and Space Administration (NASA) proposed to design an air data sensor system that is fused to the surface streamline of the aircraft so as

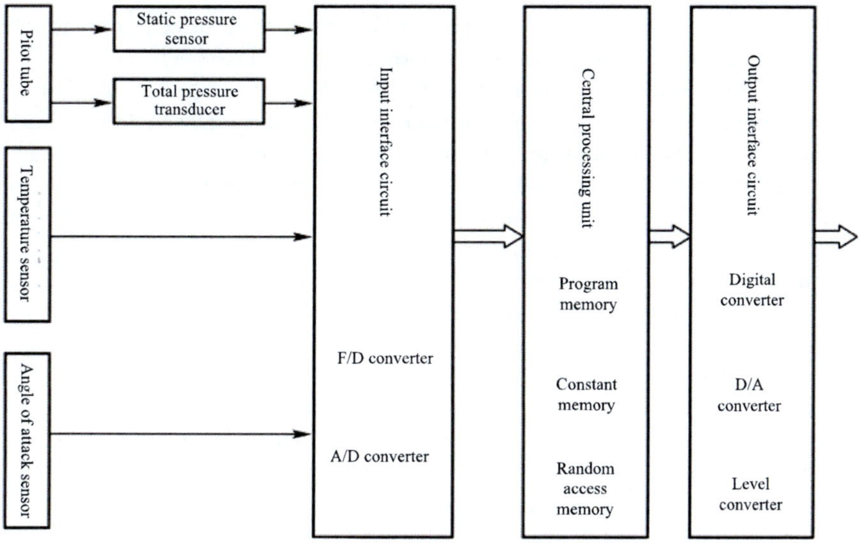

Fig. 6.7 Block diagram of principle of digital atmospheric data computer

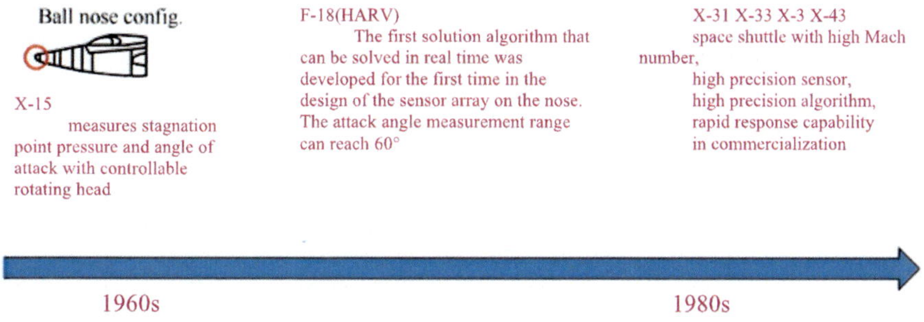

Fig. 6.8 The development of FADS in the United States

to meet the needs from the air data measurement when the space shuttle enters the atmosphere. The model of original principle was verified in the X-15 project. The model adopted a hemispherical shape structure, and a blade that can swing with the airflow is designed inside the sphere to sense the angle of incidence of the airflow. The test effect is also not ideal since the mechanical device of the system is very complicated. This idea of using mechanical devices to measure atmospheric data under hypersonic conditions was abandoned after the end of the X-15 project. The idea of using a pressure sensor array fixed on the circumference of the aircraft head to measure atmospheric data was proposed later, in the SEAD project carried out by NASA Langley Research Center. This atmospheric data measurement device does not require mechanical components, which is reversed by pressure distribution. Push flight parameters. Please refer to reference for the working principle of the FADS system of the X-15 aircraft and the arrangement of pressure taps [2].

The principle prototype developed with this idea was verified in the 1980s, but it was found that the system can also obtain better results under low Mach number conditions during the wind tunnel test, so the researchers carried out research on the application of the FADS system to other aviation fields was carried out, and a series of experimental work was carried out. These work on early stages mainly verified the relationship between the pressure at the individual pressure taps and the flight parameters, and the principle feasibility with use of pressure sensor arrays to measure atmospheric data did not develop a set of real-time mathematical algorithms to meet the need for flight control. Talents such as Stephen A. Whitmore, Timothy R. Moes and others began to devote themselves to the research until the 1990s on FADS real-time algorithms, fault detection and management, and system error estimation technologies, and conducted many laboratory simulations and wind tunnels. FADS technology gradually matured after the test and flight verification test.

The FADS technology was proposed in 1996 in the vector-driven short-range take-off and landing project cooperated by the United States and Germany, in order to solve the problem of measurement accuracy of flight parameters under high maneuverability and high maneuvering flight conditions of the X-31 aircraft. FADS technology was successfully applied to the X-31 aircraft in April 2003 after the X-31 test flight [2].

Japan conducted FADS research through the Hypersonic Flight Test (HYFLEX) program in the 1990s, as shown in Fig. 6.9.

HYFLEX was successfully launched using the J-1 rocket on February 12, 1996, but the recovery failed when the aircraft reentered. The flight trajectory, including the impact point, deviated from the design target value by more than 5 km. The rest of the flight data was transmitted back to the ground with high quality with the

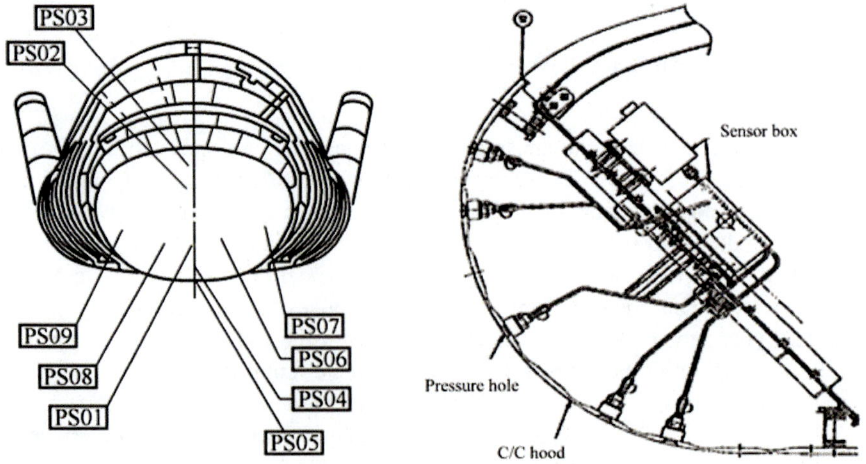

Fig. 6.9 Design program of Japan HYFLEX airborne FADS system

exception of a brief interruption caused by the Ogasawara Islands horizon. The main purpose of installing the atmospheric measurement system on HYFLEX is to obtain atmospheric data of the flying environment on the hypersonic flight trajectory, and then compare it with the results of ground tests and CFD predictions. The comparison of the flight data is to fully verify the reliability of the ground test and CFD method, and ultimately establish the "Hope" aircraft (H-II orbiter test) suitable for design tools of space reentry, suborbital speed flight and autonomous landing. It is planned to complete the mission objectives with 5 test methods.

Pressure measurement is used to verify the new concept of atmospheric measurement system (HYFLEXADS, HADS) for hypersonic flight as one of the test methods. As an embedded atmospheric data system, HADS has the same principle as the Space Shuttle Reentry Atmospheric Data System (SEADS). Flight test results show that the errors of HADS predicted angle of attack, sideslip angle and dynamic pressure are respectively 0.5°, 0.2° and 6% compared with IMU. The dynamic pressures predicted by HADS and IMU are compared with the dimensionless pressure coefficients of CFD, wind tunnel experiments and flight data although we cannot know the true value of atmospheric data. The results predicted by HADS seem to be more accurate than those predicted by IMU. The above conclusions are also verified by the test results of the NAL supersonic wind tunnel experiment [38].

The atmospheric data measurement system used by the X-43A aircraft also adopted FADS technology in the "Hyper-X" polar supersonic flight project carried out by NASA at Edward Air Force Base in California on November 16, 2004. The highest Mach number of the aircraft is close to 9.8. NASA performed flight verification on the application of FADS system with use of X-43A aircraft in the pointed wedge precursor aircraft. As a hypersonic vehicle with the main purpose of verifying the technology of the scrubber engine, X-43A aircraft replying on the INS alone cannot meet the test accuracy requirements in view of the high sensitivity of the scrubber engine inlet to the angle of attack. Another main task of the aircraft is to verify the operation stability and reliability of the FADS system for this purpose, so that the FADS system can provide a higher precision angle of attack to meet the high accuracy of the angle of attack required for the test task of the combustion engine demand. X-43A aircraft in NASA test flight is adopted to verify the stability and reliability of the FADS system in two missions with Ma = 7.0 and Ma = 10.0. The X-43A is a shape with a pointed wedge precursor for the distribution of the FADS system on the X-43A. The FADS system module is installed at the front end of the aircraft, so as to make the pressure measurement pipeline of the pressure sensor as short as possible to reduce the impact from the error of aerodynamic delay on the accuracy of angle of attack measurement [14].

The "Mars Science Laboratory" (Mars Science Laboratory, MSL) entry vehicle (EV) successfully entered the Martian atmosphere and landed safely on the Gale Crate of planet on August 5, 2012. MSL will carry the largest and most powerful Mars rover ever compared with other Mars landers and rover and MSL has the largest entry mass and the largest spherical-conical aerodynamic shell, which produces the largest hypersonic lift-to-drag ratio. It uses the largest disc slit with supersonic parachute for deceleration and carries the scientific payload with the largest volume and the largest mass. The MSL is equipped with a specially designed instrument system to measure the aerodynamic force and aerodynamic heat environment that returns to the cabin during atmospheric entry. This instrument system is called MSL Entry, Descent, and Landing Instrumentation (MEDLI). MEDLI also consists of three subsystems. The first is the MEDLI Integrated Sensor Plugs (MSP). The MISP device is a heat resistance and decay sensor measurement system that provides pneumatic heat measurement of the performance of the protective layer of the returner. The second is Mars Entry Atmospheric Data System (MEADS). The third is Sensor Supplementary Electric (SSE). SSE is an electrical box that provides signal conditions and data acquisition systems for MISP and MEADS. The MEADS consist of 7 pressure sensors with embedded pressure taps connected to the heat protection layer of the return body precursor to measure the pressure distribution.

MEDLI sensors can measure and obtain key data for reconstruction of the entry trajectory data, identification of aerodynamics, atmospheric data, heat protection systems, and testing and verification of ground system programs. The primary scientific target of MEADS is to reconstruct the atmospheric data variables separately with use of the measured pressure data when the flow pressure into the tank is greater than 850 Pa. The second target is to measure the free incoming density and atmospheric wind speed, which can be used in conjunction with airborne inertial measurement unit (Inertial Measurement Unit, IMU) data. These measurements can improve reconstruction capabilities and performance analysis of MSL trajectory data. It can ensure that aerodynamic data is separated from atmospheric data. The 3b error of the design measurement target is: the Mach number error is 0.1, and the measurement errors of the angle of attack and the sideslip angle are 0.5, the relative error of dynamic pressure is within 2%. The maximum measurement Mach number of MEADS is close to 30, which is the maximum value of the Mach number measurement of the embedded atmospheric sensor measurement system so far. For more information of MSL embedded atmospheric data sensor system, please see reference [11].

France has been committed to the research and development of air-breathing hypersonic vehicles for nearly 20 years. France formulated the National Hypersonic Research and Technology (PREPHA) program in 1992 under the leadership of the Ministry of National Defense and other companies of organizations. The PREPHA program lasted 6 years, in which it developed the Chamois scramjet and repeated tests at a Mach number of 6. The French Aeronautics and Astronautics Research Institute (ONERA) and MB-DA Company began to implement a flight test plan called LEA in January 2003, which is aimed to improve the design, development and test capabilities of France aspirated hypersonic aircraft, establish effective research and development methods include methods for research on digital simulation and ground test, and development of mathematics and test tools. These methods can be used to develop and verify aircraft and conduct flight tests.

The French LEA aircraft and its FADS system is shown in Fig. 6.10. However, FADS design and analysis data published less and its specific performance cannot be understood. EADS, a European defense company, has carried out a lot of research on the FADS system, and has now developed mature products. EADS has produced an integrated FADS system [37].

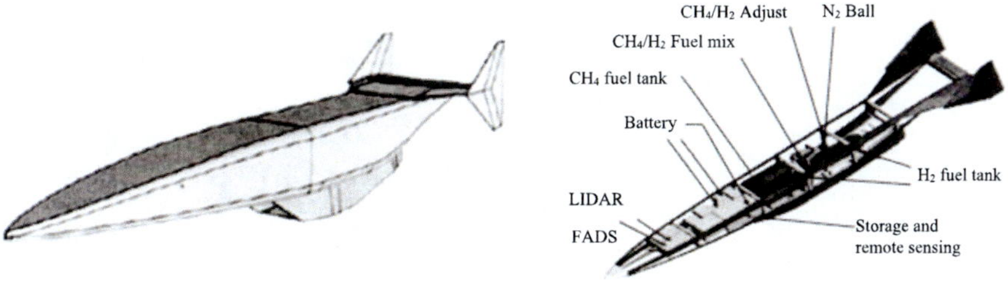

Fig. 6.10 French LEA aircraft and FADS system

Exploratory research has also been carried out on FADS technology in China. The research group of Professor Lu Yu ping in Nanjing University of Aeronautics and Astronautics has conducted a series of studies on FADS theory in terms of algorithms, calibration, system errors, calibration, FADS and INS combination, and fault detection [39–49]. Li Qichang from China Aerodynamics Research and Development Center, carried out analysis on the feasibility of a three-point solution method for embedded atmospheric data [50]· Li Qingdong from Northwestern Polytechnical University discussed FADS rapid intelligent fault detection and diagnosis technology [51–53]; Li Dongsheng, Fu Wenxing and others carried out research on FADS-based solution of the angle of attack of the aircraft [54]. Jiang Cheng, Ni Shihong, Zhang Zonglin and others from Air Force Engineering University carried out neural network algorithm simulation of embedded atmospheric data system [55].

Wang Yan and Zheng Wei from Shenyang Aircraft Design Institute conducted a preliminary study of distributed embedded atmospheric data system algorithms [56]. The configuration description of the FADS system is given for a certain type of aircraft, and the algorithm design of the FADS is carried out. The algorithm established the relationship between the pressure at each pressure measurement point and the basic atmospheric parameters with a non-physical mapping method.

The China Aerospace Aero-space Technology Research Institute has carried out research on engineering application in the areas of FADS theory, wind tunnel verification and flight test testing, etc. [37–65], and carried out a comprehensive research on the supersonic FADS system, including mechanism analysis, algorithm design, failure diagnosis and fault tolerance, system integration technology, wind tunnel calibration technology and flight test technology [57].

The first research on the embedded atmospheric data sensing system (FADS) technology for supersonic ($Ma > 3.5$) flight tests was conducted with use of the mature low-cost rocket platform-"Starry Sky 1" in China. A FADS pressure measurement system is installed in the aerodynamic load compartment of the rocket head. The pressure data of the pressure measurement hole is obtained by remote measurement as the input of the FADS algorithm calculation, which calculates the flight trajectory static pressure, Mach number, angle of attack and side slip angle. The average relative error of static pressure is about 4.1% compared with the radar measurement results. The maximum relative error is 9.2%. The average relative error of Mach number is 2.7%, and the maximum relative error is 9.3%. The theoretical ballistic results

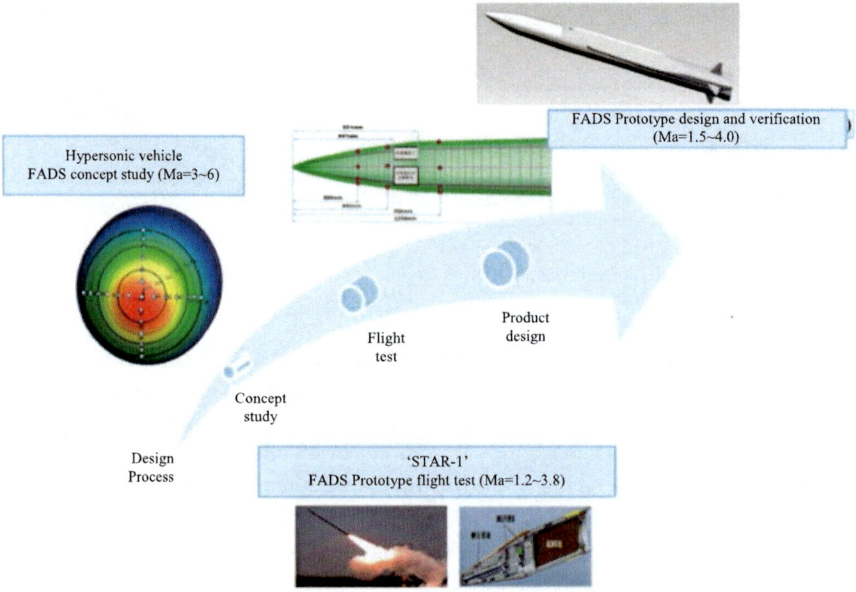

Fig. 6.11 FADS technology development process

are close. The preliminary exploration on FADS technology through low-cost flight tests has laid a technical foundation for the next step to develop mature FADS system products [58] (Fig. 6.11).

The research group of Professor Lu Yuping in Nanjing University of Aeronautics and Astronautics has conducted a series of studies on FADS theory in terms of algorithms, calibration, system errors, calibration, FADS and INS combination, and fault detection [39–49]. Li Qichang from China Aerodynamics Research and Development Center, carried out analysis on the feasibility of a three-point solution method for embedded atmospheric data [50]. Li Qingdong from Northwestern Polytechnical University discussed FADS rapid intelligent fault detection and diagnosis technology [51–53]; Li Dongsheng, Fu Wenxing and others carried out research on FADS-based solution of the angle of attack of the aircraft [54]. Jiang Cheng, Ni Shihong, Zhang Zonglin and others from Air Force Engineering University carried out neural network algorithm simulation of embedded atmospheric data system [55].

China Aerospace Aero-space Technology Research Institute has developed a supersonic embedded atmospheric sensing system CA-FADS with independent intellectual property rights in order to meet the needs of the rapid development of the domestic aerospace market, as shown in Fig. 6.12. The first use of supersonic wind in China Carry out the experimental research on the prototype of the embedded atmospheric data sensing system of the aircraft. The CA-FADS system is designed for the contour of the head of a typical supersonic vehicle. The flying Mach number is 1.5–4.5. The flying altitude is 0~25 km. The angle of attack and sideslip angle

Fig. 6.12 CA-FADS system

are $-10°$ to $10°$. It can meet the control range of the flight envelope of most supersonic aircraft. The CA-FADS system mainly consists of the pressure measurement module, the solver module and the solution algorithm model. The algorithm has active fault tolerance using a ten-shaped array layout of 9 pressure measurement points. The number of points for autonomous fault detection is 3, and the number of fault-tolerant points is 2. The CA-FADS system completed the real-time solution test of the principle prototype in a 1.2 m × 1.2 m supersonic wind tunnel. It is shown in the test results that the relative error of static pressure measurement is less than 7%, the measurement error of Mach number is less than 0.1, and the measurement errors of angle of attack and sideslip angle are less than 1 (Table 6.1) [57].

The comparison between the real-time measurement results of the CA-FADS system and the wind tunnel system with $Ma = 4$ and $\beta = 0°$ is shown in Fig. 6.13. CA-FADS system product design and test verification process are shown in Fig. 6.14. A test improvement program is also proposed in order to further improve the measurement accuracy of the system. It is found after summarizing and analyzing the test results and the factors influencing the wind tunnel test that the supersonic wind tunnel test section has the influence of step wave. The corresponding strut is designed to adjust the model position by analyzing the position of step wave of different Mach number, avoiding the influence of step wave as far as possible, and the test position with the optimal flow field quality is obtained. The research on wind tunnel improvement test verification and flight test deployment test have been completed.

Table 6.1 Statistics on error of the first test calibration results

Experimental Status	Static Pressure Error/Pa	Relative (%) Error of Static Pressure	Mach Number Error	Angle of Attack Error / (°)	Side slip Angle Error / (°)
Ma = 2.5	W970	W6.0	<0.10	<1.0	<1.0
Ma = 3.0	<400	W4.4	<0.10	<1.0	<1.0
Ma = 3.5	<300	W3.3	<0.10	<1.0	<1.0
Ma = 4.0	<300	W6.9	<0.10	<1.0	<1.0

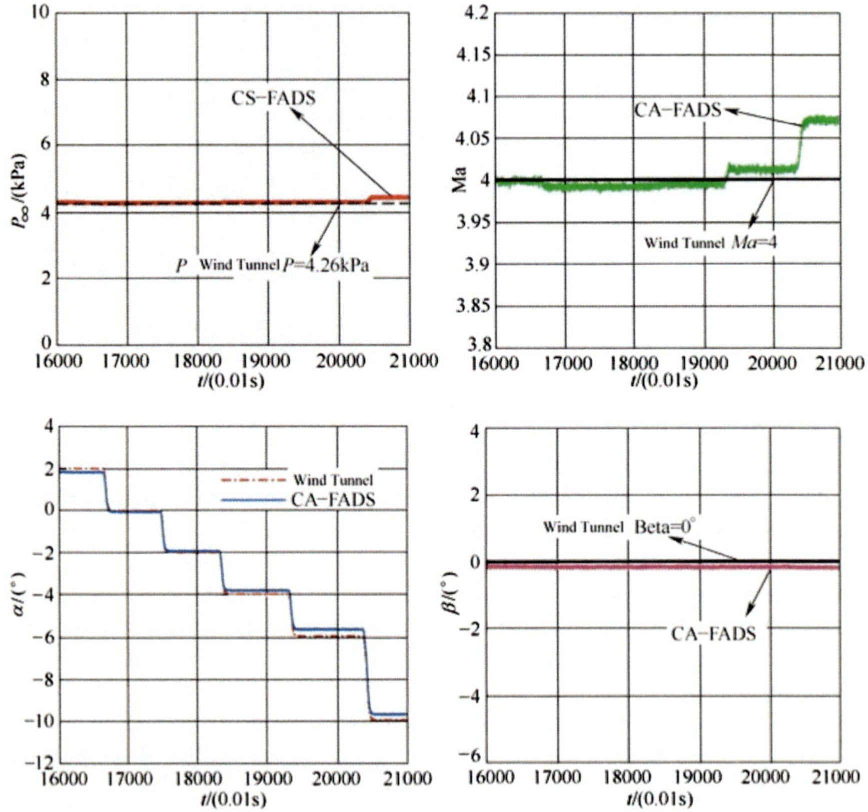

Fig. 6.13 Comparison of FADS measurement and wind tunnel measurement result when $Ma = 4$, $\beta = 0°$

FADS system improvement test is shown in Fig. 6.15. The calibration of the FADS system requires a high-quality flow field. It will cause great interference to the calibration of the FADS system due to the step wave generated at the docking position of the nozzle box of the wind tunnel, and there are differences in the position of the step wave of different Mach numbers. Therefore, it is required to analyze the position of the step wave through flow field calibration, install the model by reasonably designing the length of the strut to avoid the influence of the step, and make the model at a uniform position in the flow field to perform the calibration test of the FADS system to accurately calibrate the measurement error of the FADS system. The verification analysis on the FD-12 wind tunnel entering and passing through the flow field shows that the use of short struts (305 mm) at $Ma = 2$ and long struts (1320 mm) at $Ma = 2.5$–4.0 can effectively avoid step waves effectively and the uniform flow field can be obtained at the model installation position.

The calibration results for the ground wind tunnel test with the FADS system installed with the long pole model are: The static pressure error is less than or equal

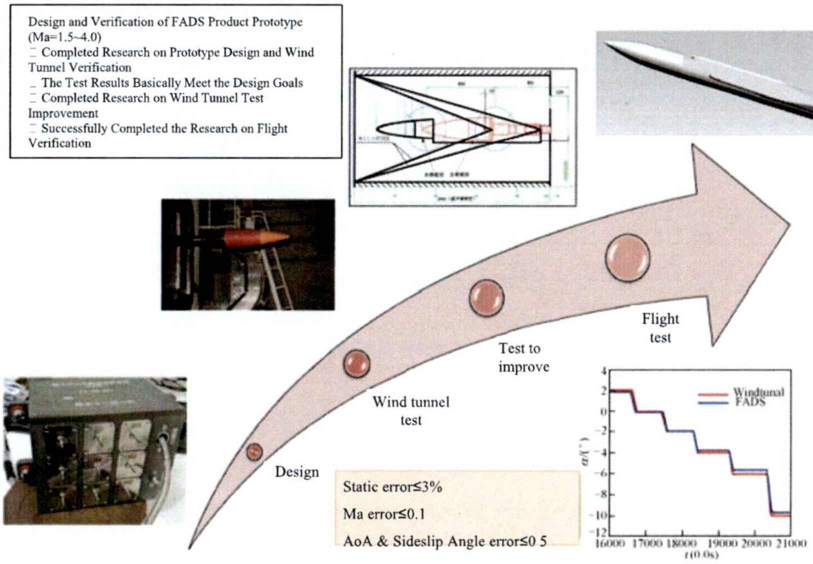

Fig. 6.14 CA-FADS system product development process

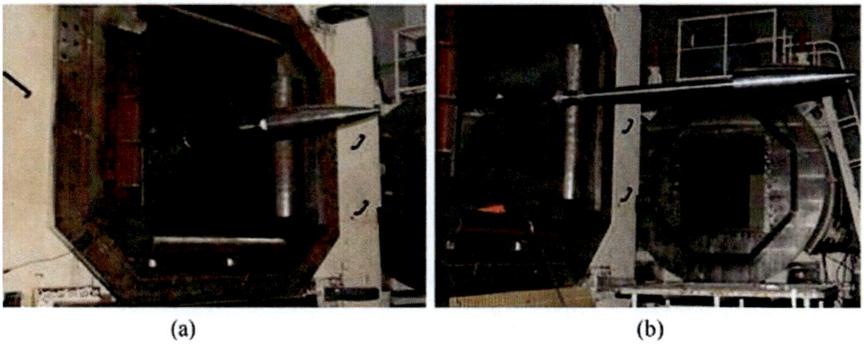

Fig. 6.15 Step wave position and model installation. **a** $Ma = 2$ step wave position and short pole model installation; **b** $Ma = 3.0$–4.0 step wave position

to 490 Pa (3%). The Mach number error is less than or equal to 0.1 and the angle of attack and side slip angle errors are less than or equal to 0.5. See Table 6.2. The wind tunnel test calibration results indicate that the design of surface FADS system meets the technical specifications. The measurement error is reduced significantly compared with the first test.

The CA-FADS system successfully completed the pilot flight test. The deviation statistics of the measured data in each segment of the flight is shown in Table 6.3, which divided trajectory into 4 segments of boost, climb, cruise and depression. It can be obtained from Table 6.3 that the measurement result deviation of the FADS system

Table 6.2 Results of FADS system wind tunnel secondary test calibration

Mach number	Static pressure error/Pa	Relative error of static pressure (%)	Mach number error	Angle of attack error/(°)	Side slip angle error/(°)
2.0	W490	W1.9	W0.05	W0.5	W0.5
2.5	W400	W2.4	W0.05	W0.5	W0.5
3.0	W200	W2.4	W0.05	W0.5	W0.5
3.5	W200	W3.0	W0.10	W0.5	W0.5

Table 6.3 Deviation of FADS measurement results of each flight segment relative to IMU superimposed wind measurement results

Flight ballistic segment	Static pressure measurement maximum/average deviation/Pa	Mach number measurement maximum/average deviation	Maximum angle of attack measurement/average deviation/(°)	Maximum sideslip angle measurement/average deviation/(°)
1- Boost Section	370/268	0.12/0.05	2.20/0.32	2.27/0.19
2-Climbing Section	300/106	0.20/0.15	0.50/0.17	1.16/0.28
3- Cruise Stage	100/67	0.17/0.10	0.50/0.21	0.27/0.10
4- Lower Pressure Section	290/161	0.18/0.09	0.90/0.45	0.46/0.12
Statistics of All Sections	370/99	0.20/0.12	2.20/0.22	2.27/0.19

and the IMU (after superimposing wind), the static pressure, angle of attack and side slip angle are affected by the engine transition in the boost section, and the pitch vibration of the body is severe. The maximum deviation is reached and the deviations are small in segments of flight such as climb, cruise and depression. The Mach number for the deviation between the climb section and the depression section reaches the maximum, while the Mach number deviation in the cruise section is smaller. The deviations are minimized of static pressure, Mach number, angle of attack and sideslip angle in the cruising section. The maximum/average deviation of the static pressure is 100 Pa and 67 Pa respectively compared with the measurement results of the Inertial Measurement Unit (IMU) in the cruise section (The altitude of the flight trajectory obtained by IMU measurement is converted into static pressure by using the standard atmospheric parameter model, and the static pressure measurement deviation of FADS system is obtained by comparing with it) The maximum Mach number/mean deviation was 0.17 and 0.10, the maximum attack Angle/mean deviation was 0.50° and 0.21°, and the maximum sideslip angle/mean deviation was 0.27° and 0.10°, respectively, which fully verified the design and measurement accuracy of FADS system.

6.1.3 Analysis on Advantages and Disadvantages of Two Atmospheric Data Sensing Systems

The sensing components of the traditional atmospheric data system consists of relatively independent functions such as probe-type airspeed tube, wind vane type (pressure difference type) angle of attack (side slip angle) sensor. They are externally mounted on the surface of the aircraft, and the airspeed, angle of attack, and sideslip angle are measured through mechanical components. The application of traditional atmospheric data systems is still very wide at present, but the application and research show that the traditional atmospheric data systems in the following applications encountered insurmountable difficulties:

(1) It is difficult to use it at hypersonic speed. Airspeed tube and angle-of-attack (slip angle) sensors will damage the streamline design of the aircraft, bring bad aerodynamic effects and even produce unstable vortex. The airspeed head is located near the stagnation point in contact with the shock layer around the aircraft head Interaction. The stagnation kinetic energy is converted into extremely high heat energy, which is easy to cause permanent thermal damage to the airspeed head.

(2) It is difficult to use in stealth conditions. Intrusive components form a vertical plane with the surface of the aircraft. It will have an adverse effect on stealth performance since the electromagnetic reflection of the vertical plane is difficult to eliminate.

(3) It is difficult to use under special circumstances. It is difficult for the airspeed head to accurately measure the incoming air pressure when flying at a large angle of attack, and the measurement accuracy of the angle of attack sensor is limited. It is difficult to measure atmospheric parameters with enhanced real-time features during high maneuvering due to the inertial characteristics of the airspeed head and the angle of attack (slip angle) sensor.

(4) High Installation, Deployment and Maintenance Costs. The angle of attack (slip angle) sensor is not suitable for installation on unconventional outline aircraft. The airspeed head will affect the installation of radar and other devices at the head of the aircraft. The application weight is increased due to the large weight and volume of the equipment. Mechanical parts loss is high with short maintenance cycle and high cost.

(5) Limited Reliability and Single Function. The sensing components of the traditional atmospheric data system are usually external components, and the living environment is harsh, for example, the airspeed head is susceptible to moisture, freezing, blockage, etc. The angle of attack (slip angle) sensor function is monotonous, and the application is basically not scalable.

The FADS System has been Widely used in Modern Aircraft, Mainly Because it has the Following Advantages.

(1) It does not affect the aerodynamic contour of the aircraft. The FADS system has no parts outside the surface of the aircraft. All components are integrated inside the aircraft, regardless of the aerodynamic contour of the aircraft.
(2) The measurement accuracy is high. The measurement error of angle of attack and sideslip angle is 0.5. Within, the dynamic and static pressure measurement error is within 5%, even hypersonic speed and other conditions can achieve such accuracy under high angle of attack.
(3) The range of applications is wide. It can meet the needs of atmospheric data measurement under high-angle flight conditions. The entire system of FADS is embedded within the surface of the aircraft, thus avoiding the high-temperature surface of the aircraft during hypersonic flight, so that the entire system is protected. It can also reduce the transmitting area of radar and improve the stealth performance of aircraft. The FADS system has relatively high performance in a wide range of angle of attack and Mach number. In addition, FADS can also be used on some micro UAVs.
(4) There is no mechanical moving parts in the system, which is convenient for system integration and maintenance. The components of the entire system are fixed, without mechanical devices, which is convenient for miniaturization and integration of the system, and greatly reduces the probability of system failure at the same time with easy maintenance.
(5) It is convenient for redundant configuration and greatly improves system reliability. Usually three or four atmospheric data systems are required if a traditional atmospheric data system is used, in order to ensure that the system has sufficient redundancy to ensure the failure survivability of aircraft. The redundant configuration of FADS is relatively easy, which only needs to arrange a few more pressure measuring holes and then can solve multiple sets of atmospheric data.

The FADS system is the main way for the spacecraft to obtain flight atmospheric parameters as flight control input, especially for the aspirated spacecraft. Dynamic pressure and angle of attack are two key design parameters for the design of the sub-combustion/super-combustion ramjet engine of the aspirated space vehicle, which determine the intake port design of sub-combustion/super-combustion ramjet engine. It is of great significance to obtain accurate flight parameters for real-time control of the aircraft and post-flight data analysis. The performance of sub-combustion/super-combustion engines is very sensitive to the angle of attack. Accurate and real-time angle of attack measurement can ensure the best flight performance of sub-combustion/super-combustion ramjet engines. It cannot reflect the changes of the atmosphere and the interference of external conditions such as wind speed due to the impact of the installation and performance of the inertial navigation system, so it is impossible to accurately obtain the angle of attack of flight flow. The FADS system is featured by high accuracy, low cost and simple maintenance. Therefore, the use of FADS to measure atmospheric parameters such as the inflow angle of flight of the spacecraft as the input of the control system is of great significance for reducing the cost of aircraft manufacturing and improving overall performance.

6.2 Survey of the Critical Technologies

The spacecraft FADS system is a high-precision measuring instrument. Its development process is a complex system engineering, which requires close cooperation of various majors and links before it can finally develop a qualified product that meets the needs of engineering applications. Usually it is required to go through the following processes: requirements analysis, overall design, hardware system design, aerodynamic design/testing, algorithm design, integration, manufacturing, wind tunnel testing and flight test verification. Therefore, the development of FADS system requires solving the following key technical problems.

(1) FADS overall design technology;
(2) High-precision solution algorithm technology with multi-point fault tolerance;
(3) High-precision and low-delay pressure measurement technology;
(4) Solver hardware design technology;
(5) FADS system wind tunnel test and flight test technology;
(6) Test technology of FADS system in ground environment;
(7) FADS and INS/GPS data fusion technology.

6.2.1 Overall Design

The overall design study of the FADS system is carried out and the overall program of the FADS system is determined (Fig. 6.16) according to the requirements of aerospace vehicle design trajectory, measurement accuracy requirements, airborne equipment layout space and data interface, etc. The overall design includes measurement accuracy index, system solution time, location of distribution points, total number of pressure-measuring points, fault-tolerant requirements, pipeline layout of pressure measuring, data interface, data fusion and working environment conditions.

6.2.2 Algorithm Design

FADS has developed three solving algorithms at present, namely the three-point method, the least square method and the neural network method. Three-point method: accuracy is high and anti-interference performance is good, while the iterative real-time performance of the three-point method is poor. Moreover, the selection of pressure measuring holes will have great influence on the calculation results and even lead to ill condition of the system. Least square method: The most prominent advantage is

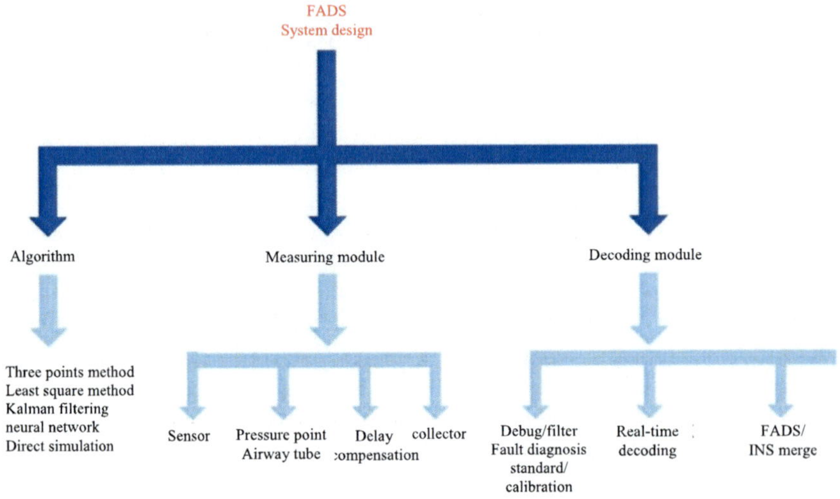

Fig. 6.16 Overall design of FADS system

that the algorithm has nothing to do with the position of the pressure measuring hole and the failure processing is relatively simple. Its shortcoming is that it is realized through iteration, so that convergence problems will take place inevitably. Neural network method: Its powerful nonlinear mapping ability is adopted to deal with the nonlinear problems of the FADS system. The results of the least squares method and the three-point method need to be corrected by appropriate methods, while the neural network method avoids this process, so it is easy to calculate in real time.

The disadvantage of this algorithm is that it requires a large amount of data to train the network, the algorithm takes a long time to prepare, and the network weights and thresholds cannot be given physical meaning.

CFD calculation is adopted to obtain the pressure distribution of the contour of the head of the aircraft as the input of the neural network sample training, and the corresponding incoming state, such as static pressure, Mach number, angle of attack and side slip angle, is used as the target training neural network for the sample. A FADS solving algorithm based on the BP neural network is established with the powerful non-linear mapping ability of the neural network (Fig. 6.17). Arrange the pressure measurement points according to the design goals, and design the redundancy and fault tolerance scheme of the system (Fig. 6.18).

6.2.3 Pressure Measurement Module Design

The measurement module mainly consists of the design of the pressure measurement pipeline and the sensor selection. The size and inner diameter of the pressure

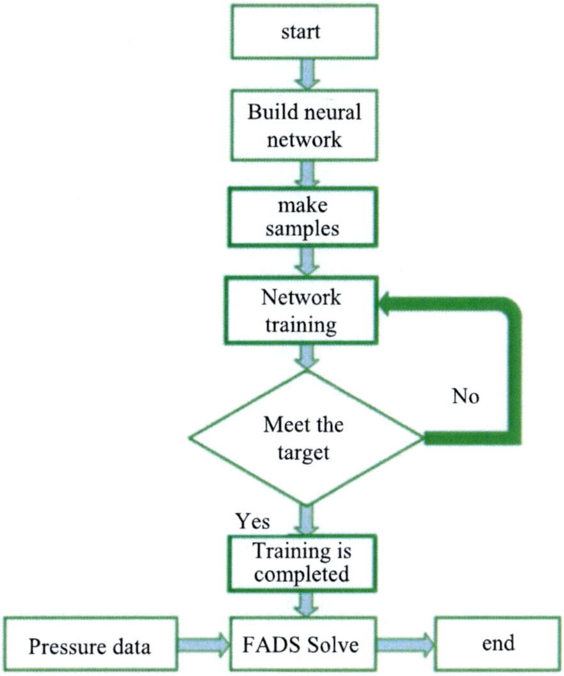

Fig. 6.17 Design process of FADS algorithm based on BP neural network

measurement pipeline are designed according to the requirements of the on-board equipment layout of the aircraft and the delay time of the pipeline. The airflow ambient temperature of the pipeline at the pressure measurement point is analyzed through trajectory simulation, and the heat protection scheme of the pipeline and the sensor is designed according to the simulation result. The sensor range and accuracy are selected based on the following three design principles by analyzing the pressure of the full trajectory measurement point at each cross-sectional position: ① the sensor does not damage under full trajectory flight conditions. ② Ensure that the FADS system range can meet the design requirements. ③ The simulation analysis is carried out according to the requirements of the system design index to determine the measurement accuracy of the sensor, and the pressure sensor is selected for reference.

There are many types of pressure sensors (Fig. 6.19), including resistance strain gauge pressure sensors, semiconductor strain gauge pressure sensors, piezoresistive pressure sensors, capacitive pressure sensors, resonant pressure sensors, piezoelectric sensors, and inductive pressure sensors.

As currently the most mature and widely used type of pressure sensors, silicon piezoresistive pressure sensors have high sensitivity, small size, good repeatability, strong overload capacity, and wide frequency response. Most of the FADS systems on models from abroad use silicon piezoresistive pressure sensors.

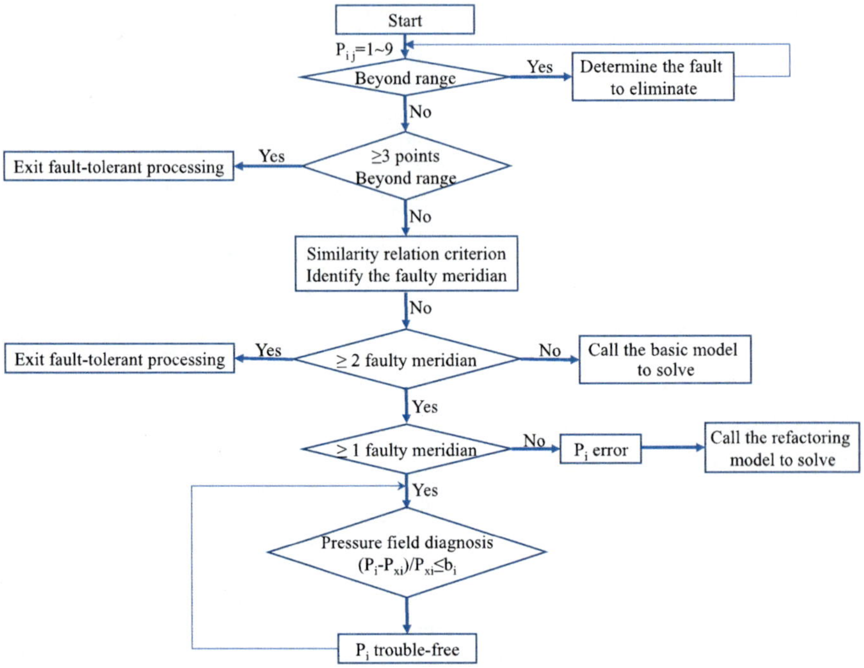

Fig. 6.18 The active fault tolerance design scheme for FADS algorithm (taking 9 pressure measurement points as an example)

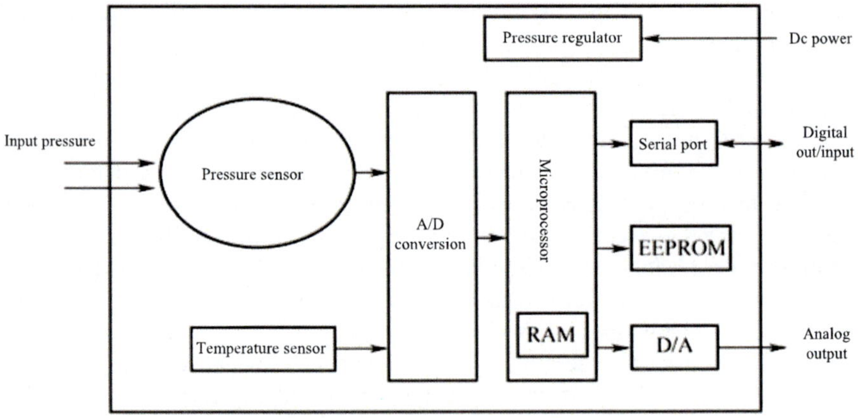

Fig. 6.19 Working principle of pressure sensor

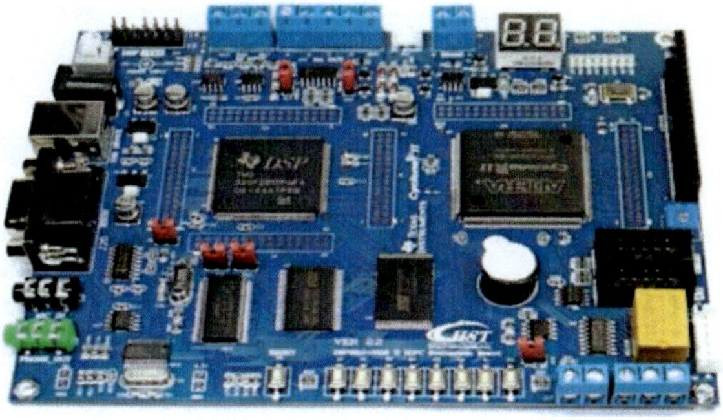

Fig. 6.20 Schematic diagram based on FPGA + DSP architecture

6.2.4 Solving Module Design

The solver adopts FPGA + DSP architecture design (Fig. 6.20), with full combination of the advantages and characteristics of each. FPGA is suitable for large data volume and high processing speed requirements, but the calculation structure is relatively simple. DSP is suitable for high-level signal algorithm processing with high operation speed and complicated control structure. The solver is used to collect pressure sensor data in real time, and accomplish the corresponding calculation on the collected data. The solver sends the collected and processed data after the solution to the telemetry system through one RS422 communication interface, sends the processed data to the control system through another RS422 communication interface, and receives and processes control commands from the control system at the same time. The data resolution time is not greater than 5 ms.

6.2.5 Wind Tunnel Test Research

A wind tunnel test verification study can be carried out on the system after completing the preliminary design of the FADS system and the integration of various components (Fig. 6.21). The typical trajectory points are selected according to the trajectory design of the aircraft to carry out the real-time calculation study of the embedded atmospheric sensor system wind tunnel test, verify the design technical indicators, improve the system design and improve the measurement accuracy through the analysis of the test results. The wind tunnel test of the FADS system is a high-precision pressure test with strict requirements on the quality of the wind tunnel flow field, model processing and installation accuracy. There is usually some interference in the flow field of the wind tunnel test, such as the influence of the angle of the airflow and the step wave generated by the nozzle docking. It must be verified and analyzed to obtain a uniform quality location for FADS system testing. The position of the step wave is calibrated by the flow field, and the support rod of the design model adjusts the position of the model to avoid the step wave. The fine adjustment mechanism of

Fig. 6.21 Wind tunnel test of FADS system installed on the model

the support rod is designed for the influence of the airflow declination, and the model is fine-tuned according to the on-site calibration to eliminate the airflow declination. In general, the installation accuracy of the pressure measurement point of the wind tunnel model by the FADS system requires an error ≤ 0.1 mm, and the model size processing accuracy ≤ 0.02 mm.

6.2.6 Test and Research of Ground Environment

The FADS system must carry out ground environmental test tests on the system after completing the wind tunnel test verification and reaching the design index, since the working environment of the FADS system is usually relatively harsh.

Mainly includes component testing and overall testing. Vibration testing, lightning strike testing, and electromagnetic compatibility testing are mainly required in component testing and overall testing, in order to analyze the performance of the pressure sensor and related electrical measurement values. Vibration test (Fig. 6.22a) is to add vibrations that may encounter vibration frequency and amplitude in the three axes of the sensor in actual flight. The mechanical characteristics of mechanical components and the sensitivity of the sensor to vibration shall be tested that whether

Fig. 6.22 Technical flight test of "Starry Sky"-1 FADS

it will be affected by these vibrations. In addition, the atmospheric data computer also needs to perform vibration tests to ensure that the vibrations will not affect the real-time data calculation. Lightning protection tests and mechanical strength tests are also required. The electromagnetic compatibility/electromagnetic interference test is adopted to check the sensitivity of FADS sensors and atmospheric data computers to electromagnetic interference such as external electromagnetic radiation. Certain measures are taken to avoid external influences when necessary [37].

6.2.7 *Flight Test Research*

It is also necessary to install the FADS system on the actual aircraft for flight test after the system completes the wind tunnel test and ground environment test (Fig. 6.22 is "Starry Sky No. 1 FADS technical flight test verification) [62] to obtain the FADS system data measured in real time, which is compared with the data of the radar system and the airborne inertial navigation system, so as to verify the measurement accuracy and reliability of the system and test the influence of the airflow on the system.

Flight test results are summarized and analyzed to improve the FADS system. The FADS system only calculated and recorded atmospheric data, but did not participate in the control system during the flight test. The pressure data obtained during the flight test is used to further correct the FADS system. The flight test research is expensive and the system is very complicated. The FADS system must be conducted at the same time as the flight test of other verification technologies on the aircraft to reduce the cost of the test.

6.2.8 *Research on FADS and INS/GPS Data Fusion*

The technical indicators have reached the overall design requirements after completing the wind tunnel test, ground environment test and flight test. FADS system products can basically be finalized, and can be added to the flight control system. The aircraft control system is also installed with other measurement systems as usual, such as inertial navigation system (INS), global satellite positioning system (GPS or China Beidou system), so it is necessary to fuse the data sources obtained by the above three devices and deliver them to the flight control system uniformly.

As an autonomous navigation sensor, the inertial navigation system currently mainly uses mechanical or laser gyros. The error is too large at long distances since the inertial device itself has a congenital offset, and it needs to rely on other navigation resources for calibration. The navigation source used for calibration is the traditional radio navigation source as usual. GPS is now used to calibrate INS since the coverage and stability of GPS signals are getting better and better, that is, GPS + INS is the primary navigation resource. On the hardware, the embedded atmospheric data system FADS is usually combined with other navigation resources since the FADS system has more advantages in measuring incoming parameters, that is, GPS + INS + FADS, which is called a combined navigation source in the domestic aerospace industry. GPS + FADS combination or INS + FADS combination navigation source can also be used in order to reduce the development cost of the system.

6.3 Summary

The FADS system is featured by high accuracy, low cost and simple maintenance. Therefore, it is of great significance to measure the flight parameters of the spacecraft as the input of the control system with use of the FADS system so as to improve the overall performance of the aircraft and reducing the development cost. As the current advanced atmospheric sensing technology, the embedded atmospheric data sensing technology not only surpasses the traditional probe technology in measurement accuracy, system reliability, and stability, but also does not affect the technical characteristics of the aerodynamic contour. It is decided that the embedded atmospheric sensing technology can be applied to the advanced aircraft with the characteristics of high speed, high mobility and stealth. Therefore, this technology is a key solution in the development of future hypersonic vehicles, air-to-air shuttles, and new fighters. The overall design, system algorithm, pressure point layout, air induction structure, pressure sensor, solver, ground test, flight test loading and data fusion are several aspects that need to be studied in the development of the embedded atmospheric data sensing system. Especially for high-speed aircraft, the problems of strong shock waves, high temperature, and strong vibration in harsh flight environments are difficult and important points for development of system.

References

1. Nan B, Zhaofeng S, Jingchun Y (2010) Research on embedded atmospheric data sensing technology. Flying Missile 8:79–85
2. Peng W, Xin J, Weimin Z (2013) Application of FADS system in various types of aircraft. Flying Missile 2:75–79
3. Ruiheng W, Shouduo Z, Wei Y (2008) Development status of embedded atmospheric data sensing technology. Electro-Optics Control 15(8):53–56
4. Chengjun Z (2005) Research on embedded atmospheric data sensing system and algorithm. Nanjing University of Aeronautics and Astronautics, Nanjing
5. Xigao F (2007) Research on technology and application of embedded atmospheric data sensing system. Nanjing University of Aeronautics and Astronautics, Nanjing
6. Xiuyi S (2007) Research on algorithm and application of embedded atmospheric data sensing system. Nanjing University of Aeronautics and Astronautics, Nanjing
7. Lei Z (2010) Research on fault detection and processing algorithm of embedded atmospheric data sensing system. Nanjing University of Aeronautics and Astronautics, Nanjing
8. Dibo X (2010) Research on the algorithm and key technology of embedded atmospheric data sensing system. Nanjing University of Aeronautics and Astronautics, Nanjing

9. Guoqing S (2012) Error analysis and elimination method research of embedded atmospheric data sensing system. Nanjing University of Aeronautics and Astronautics, Nanjing
10. Jun X, Zonglin Z, Huicong L et al (2002) New equipment and new technologies for aviation control engineering. Aviation Industry Press, Beijing
11. Karlgaard CD, Kutty P (2013) Mars entry atmospheric data system trajectory reconstruction algorithms and flight results. AIAA-2013-28 2:12–16
12. Joelc E, Stephen AW (2007) Reentry airdata system for a sub-orbital spacecraft based on X-34 design. AIAA-2007-1200 23:15–16
13. Jost M, Schwegmann F, Kohler T (2004) Flush air data system and advanced airdata system for the aerospace industry. AIAA-2004-5028 34:12–16
14. Ethan B, Joseph WP, Mark CD (2008) X-43A flush airdata sensing system flight test results. AIAA 2008–657(5):46–49
15. Joelc E, Stephen AW (2008) Simulation of a flush air-data system for transatmospheric vehicles. J Spacecraft Rocket 45(4):716–773
16. Judy LS (1980) Comparison of predicted and experimental real-gas pressure distributions on space shuttle orbiter nose for shuttle entry air data system. NASA-TP-1627
17. Terry JL, Paul M (1980) Use of nose cap and fuselage pressure orifices for determination of air data for space shuttle orbiter below supersonic speeds. NASA-TP-1643
18. Terry JL, Stuartg F, Aulm P (1980) Wind tunnel investigation of an all flush orifice air data system for a large subsonic aircraft. NASA-TP-1642
19. Terry JL, Paul M (1981) Subsonic tests of an all flush pressure orifice air data system. NASA-TP-1871
20. Stephen AW, Timothy RM, Terry JL (1990) Preliminary results from a subsonic high angle-of-attack flush air data sensing (HI-FADS) system design calibration and flight test evaluation. NASA-TM-101713
21. Stephen AW (1999) Development of a pneumatic high -angle-of-attack flush airdata sensing (HI-FADS) system. NASA-TM-104241
22. Timothy RM, Stephen AW, Frank L (1993) Flight and wind-tunnel calibrations of a flush air data sensor at high angles of attack and sideslip and at super-sonic mach numbers. AIAA-93-1017
23. Stephen AW, Timothy RM, Terry JL (1992) High angle-of-attack flush airdata sensing system. J Aircraft 29(5):915–919
24. Stephen AW, Cobleigh B, Edward AH (1998) Design and calibration of the X-33 flush airdata sensing (FADS) system. AIAA-98-0201
25. Stephen AW, Roy JD, John MF (1996) In-flight demonstration of a real-time flush airdata sensing system. J Aircraft 33(5):970–977
26. Thomas JR, Stephen AW (1999) Fault-tolerantneural network algorithm for flush air data sensing. J Aircraft 36(3):541–549
27. Brent RC, Stephen AW, Edward AH (1999) Flush air data sensing (FADS) system calibration procedures and results for blunt forebodies. AIAA-99-4816 5:15–16
28. Srinatha CV, Sastry KS, Raman B et al (2001) Failure management scheme for use in a flush air data system. Aircr Des 4:151–162
29. Susanne W (2002) Comparing three algorithms for modeling flush air data systems. AIAA 2002–0535(5):23–26
30. Detlef RO, Thomas G (2004) X-31 vector system identification-approach and results. AIAA-2004-4830 6:45–54
31. Jost M, Schwegmann F (2004) Flush air data system in an advanced air data system for the aerospace industry. AIAA-2004-5028 4:46–49
32. Joel CE, Stephen AW (2007) Reentry air data system for a sub-orbital spacecraft based on X-34 design. AIAA-2007-1200 3:15–17
33. Joel CE, Stephen AW (2008) Simulation of a flush air-data system for transatmospheric vehicles. J Spacecraft Rocket 45(4):716–732
34. Mark CD, Joseph WP, White JT (2000) Development of a flush air data sensing system on a sharp nosed vehicle for flight at mach 3 to 8. AIAA-2000-0504 2:56–60

35. Ethan B, Joseph WP, Mark CD (2008) X-43A flush air data sensing system flight test results. AIAA 2008–6570(16):63–67
36. Ian AJ, Peter AJ, Takayuki S (1998) A study of flush air data system calibration using numerical simulation. AIAA-98-1606 5:16–19
37. Jost M, Schwegmann F (2004) Flush air data system-an advanced air data system for the aerospace industry. AIAA Guidance, Navigation, and control Conference and Exhibit. Rhode Island: AIAA 2004-5028 2004:1–15
38. Bin Z, Shenglin Y (2006) Neural network algorithm for embedded flight parameter sensing system. J Aeronautics 27(2):294–298
39. Chengjun Z, Lu Y, Lu G (2006) Application of BP network in embedded atmospheric data sensing system. Measurement Control Technol 25(6):9–12
40. Jianchao F, Chengjun Z, Feng C (2006) Embedded atmospheric data sensing system and its BP network correction algorithm. Comput Measurement Control 14(4):541–544
41. Shouduo Z, Yuping L, Wei Y (2007) Research on FADS system fault detection and management technology based on X2 inspection. Comput Measurement Control 15(11):1449–1454
42. Xigao F, Yuping L (2008) Research on the solving algorithm of embedded atmospheric data sensing system. Comput Measurement Control 16(3):398–400
43. Xigao F, Yuping L, Shouduo Z (2008) Application research of FADS system in flying wing layout aircraft. Aircr Des 28:43–46
44. Xigao F, Yuping L, Bo L Research on a FADS system scheme installed on the leading edge of aircraft wing. Sens Microsyst 27(3):55–57
45. Xiuyi S, Yuping L (2007) Design and research of pressure sensor in embedded atmospheric data sensing system. Theory Pract 27(5):8–19
46. Xiuyi S, Yuping L (2008) Embedded atmospheric data sensing system and its correction. J Appl Sci 26(3):301–306
47. Yu Y, Yuping L, Zaigui W (2009) Combined filter technology in embedded atmospheric data sensing system. Sens Microsyst 28(5):117–120
48. Guoqing S, Yuping L, Zhihui X (2012) Error analysis of embedded atmospheric data sensing system. Sens Microsyst 31(6):62–65
49. Lei Z, Yuping L (2012) FADS system and algorithm based on RBF neural network. Aircraft Des 2012(3):3–6
50. Qichang L (2014) Preliminary study on three-point solution method of embedded atmospheric data. J Aerodyn 32(03):360–363
51. Qingdong L, Lulu C, Xiaogong Z et al FADS fast intelligent fault detection and diagnosis technology. Syst Eng Electron Technol 31(10):2544–2546
52. Yangming G, Qingdong L, Xiaobin C et al FADS sensor fault detection method based on parity equation. Aeronaut Comput Technol 40(2):98–100
53. Qingdong L, Wei W, Zhang R (2008) Research on the redundant configuration of FADS pressure sensor. Comput Simul 25(11):48–51
54. Dongsheng L, Wenxing F (2012) Research on FADS-based method for solving the angle of attack of aircraft. Sci Technol Eng 12(3):573–577
55. Cheng J, Shihong N, Zonglin Z et al (2008) Artificial neural network algorithm simulation of embedded atmospheric data system. J Syst Simul 20(9):2288–2291
56. Yan W, Wei Z (2008) Preliminary research on distributed embedded atmospheric data system algorithm. Aircr Des 28(6):5–26
57. Guangqiang C, Wuyue L, Xiuxin D et al (2016) Design of FADS system for air-to-air missile. Sci China Technol Sci 46:1193–1206
58. Guangqiang C, Guidong W, Bingyan C et al (2015) FADS technology research of low-cost flight test platform. J Astronautics 36(10):747–754
59. Guangqiang C, Guidong W, Bingyan C et al (2015) FADS algorithm research of hypersonic vehicle. Aircr Des 35(6):1–7
60. Rong W, Hongjun Z, Guidong W et al (2016) Multidisciplinary integrated design of the shape of aspirated air-to-air missile. J Aeronautics 37(1):207–215

61. Peng W, Qiuhong L, Yuansi H et al (2015) Research on the accuracy of FADS-a for pointed wedge precursor aircraft. Sci China Phys Mech Astron 45:124709-1-11
62. Peng W, Xin J, Weimin Z (2013) Application of FADS system in hypersonic vehicle with sharp wedge precursor. Sci China Phys Mech Astron 43:1105–1110
63. Peng W, Yuansi H, Xin J (2016) The effect of stagnant point pressure on the accuracy of neural network algorithms in the FADS system of sharp wedge precursor vehicles. Astronautics J 37(9):1072–1079
64. Peng W, Xin J, Weimin Z (2016) Calibration of FADS system for blunt body. Exp Fluid Mech 30(2):97–102
65. Yongming Q, Chun Z, Jingang D (2015) Experimental study on wind tunnel calibration for embedded atmospheric data sensing system. J Aerodyn 33(4):488–492

Chapter 7
Advanced Wind Tunnel Measurement Technology

7.1 Introduction

The technology on non-contact measurement of wind tunnel is a new and advanced optical measurement technology. It has the advantages of non-contact, unchanged model contour, high precision, high resolution and high measurement efficiency, compared with traditional technology of sensor point measurement. The development of aircraft design requires wind tunnel tests to provide more accurate and comprehensive data. The technology on non-contact measurement of wind tunnel can not only improve data accuracy, but also improve test efficiency. It is an important development direction in the field of wind tunnel testing in the future.

The part below is a detailed introduction to the research and development history and engineering application cases of the existing advanced technology on non-contact measurement of wind tunnel in China, including Phosphor Thermography technology, particle image velocimetry technology. This thesis introduces the development and research status of these new technologies at home and abroad, measurement principles and methods, key technologies and applications, and looks forward to the future development of non-contact measurement technology in wind tunnel.

7.2 Phosphor Thermography

As a wind tunnel optical surface measurement technology, Phosphor Thermography applies temperature-sensitive rare earth elements as luminescent materials, and uses a luminescent coating sprayed on the surface of the model to emit the characteristic that the intensity of visible light changes with temperature under a certain intensity of ultraviolet light. Test model is used to predict the surface heat environment in the wind tunnel, which has the advantages of high precision, non-contact full-field fine measurement, wide temperature measurement range, and reusability.

M. Zhao, *Prediction and Validation Technologies of Aerodynamic Force and Heat for Hypersonic Vehicle Design*, Springer Aerospace Technology,
https://doi.org/10.1007/978-981-33-6526-1_7

7.2.1 Development and Research Status of Phosphor Thermography Technology from Home and Abroad

The research on phosphorescent luminescence was started in the late 1940s [1], and has since been widely used in commercial and home lighting and improved cathode ray tubes. It was discovered at that time that the growth in ambient temperature would change the luminescence of the material, which affected the performance of related parts, and prompted its gradual application in the field of heat measurement. Czysz et al. [2] realized the application of phosphorescence thermography technology in wind tunnel at Arnold Engineering Development Center (AEDC) in the United States. The technology has gradually matured and is still in use today after decades of improvement and perfection. Borella [3] and others developed a phosphorescent temperature measuring agent with a temperature upper limit of 1200 °C. Oak Ridge National Laboratory (0RNL) in Tennessee, USA has developed a dynamic high-temperature Phosphor Thermography technology [4] with a response time of milliseconds. The first time that the Phosphor Thermography Technology was applied to the pulsed wind tunnel for surface temperature and heat flow measurement was accomplished by McDonnell Company [5]. The operating time of the pulsed wind tunnel was 50 ms, and then Cornell Aeronautical Laboratory (CAL) took this technique in application to a shock wave wind tunnel with a running time of only 5 ms, and achieved relatively ideal results [6].

The two-color Phosphor Thermography Technology appeared in the 1980s, and it developed rapidly in the Western countries, especially in the United States.

This technology is based on the monochromatic Phosphor Thermography system, which makes the luminescent material emit two different spectrums of light (usually red light and blue light) at the same time. The two cameras record the intensity changes of the response light. And the temperature distribution and heat flow distribution of the whole model are calculated by the relation between light intensity ratio and temperature. The two-color phosphorescence technology avoids the system error caused by the uneven excitation light intensity and the background noise of the acquisition device compared with the monochromatic Phosphor Thermography technology.

The Phosphor Thermography Technology was listed as the standard technology for aerodynamic heating measurement of aircraft surfaces by the Hypersonic Facility Center (HFC) of NASA Langley Center in the early 1990s [7, 8]. This technology has now become one of the basic quantitative testing methods for all aerodynamic thermal tests in Langley Center after more than 20 years of development, and has participated in the measurement of aerodynamic heat tests such as X-33.X-34.X-38. More than 70 ceramic models of were subjected to aero-thermal tests during the development of X-33, and the number of blows exceeded 1100 times. It not only provided full-surface aerodynamic heat data and conducted transition analysis, but also carried out measurement and analysis on the aerodynamic heat environment of key parts of many aircraft designs (such as the leading edge of the tail, the engine nozzle, the inclined aileron surface), and the windward surface laminar flow aerodynamic heat data obtained from the wind tunnel was extrapolated to obtain the aircraft surface temperature data under flight conditions [9–11], which plays an important role in the design of the whole machine thermal protection system.

Horvat et al. [12] made use of the Phosphor Thermography Technology to research on the influence of the groove on the windward surface on the boundary layer transition of the space shuttle, and formed a boundary layer transition database. The experimental results of the influence of grooves on boundary layer transition can clearly show the boundary layer transition phenomenon on the surface by phosphorescence thermography technology.

Michelle and Scott [13] et al. researched on the influence of different wing leading edge radius and inclination angle on peak heat flow under shock wave/shock wave interference in wind tunnel with Mach Number 10 in Langley Center.

Dennis [14] researched on the transition problem of the conical boundary layer with temperature-sensitive paint technology in the static wind tunnel of Purdue University ($Ma = 6$), and successfully captured the lateral stripe structure of the cone under the angle of attack.

Derek [15] researched the influence of heat flow on the bottom surface of the CEV with different roughness units on the bottom surface of the CEV in the Mach 6 conventional wind tunnel by using of Phosphor Thermography Technology at the Langley Research Center. The studied rough elements, including pits and bumps, accurately capture the heat flow distribution in the interference region of rough elements under different Reynolds Numbers.

Inna and Joseph [16] applied the Phosphor Thermography Technology in the CEV outsole heat environment measurement test, and obtained high-precision test results in a large area. However, a larger measurement error occurs due to the high heat flow value of the shoulder and the large local curvature radius.

China Academy of Aerospace Aerodynamics has carried out pulse wind tunnel test research on flat plate-protrusion model, compressed corner model and complex shape aircraft model with phosphorescence thermography technology. The heat flow distribution in the bump interference region (Fig. 7.1), the heat flow distribution in the compression corner high heat flow region caused by shock wave/boundary layer interference (Fig. 7.2) and the global heat flow distribution of complex shape aircraft [17–19] (Fig. 7.3) are obtained at Mach number 8.

It can be seen based on the current status of research and development at home and abroad that the application research of thermography technology in the world has become the research direction of Master in the field of hypersonics. High-precision non-contact heat flow measurement technology has become the technical dominating terrain features of hypersonic aircraft heat environment test research, and the development of hypersonic weapon equipment is of great significance.

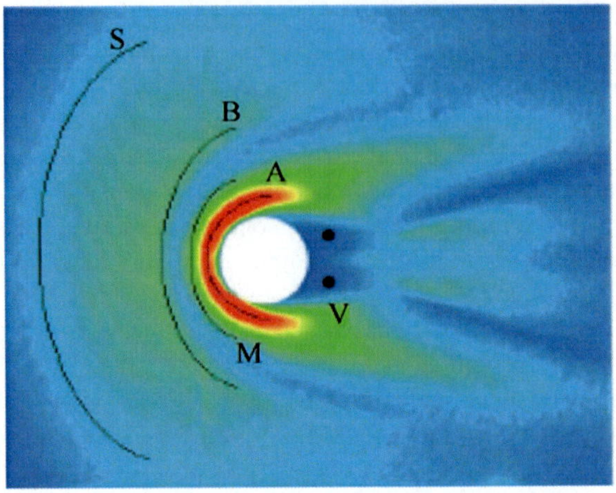

Fig. 7.1 Heat flow distribution in the interference zone of the plate-protrusion

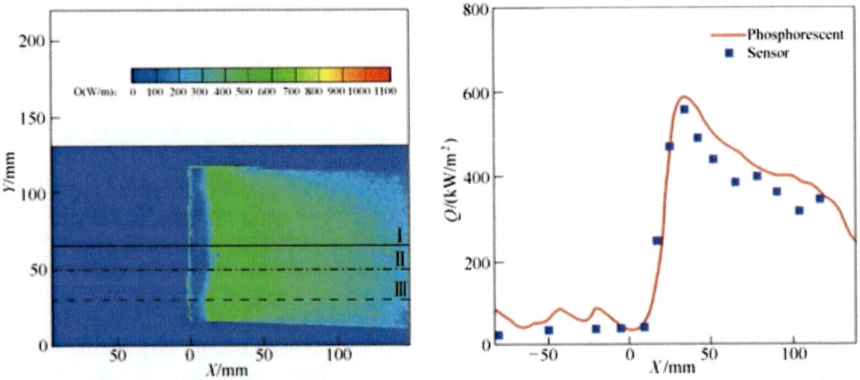

Fig. 7.2 Heat flow distribution at compressed corner

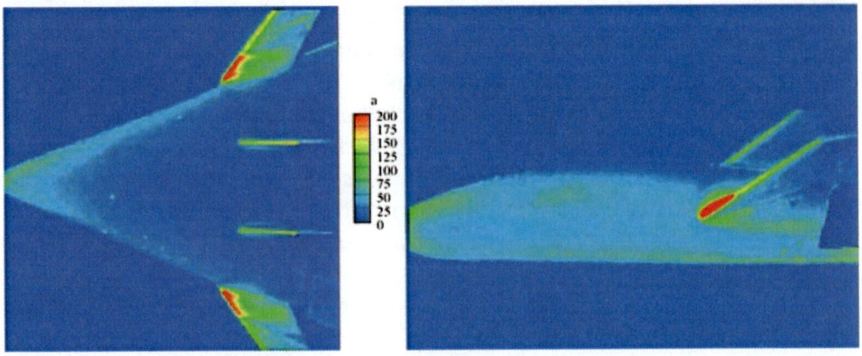

Fig. 7.3 Whole machine heat flow distribution of hypersonic aircraft

7.2.2 Measuring Principle of Phosphor Thermography Technology

Phosphor Thermography Technology is an efficient, accurate and high-precision method for full-field heat flow measurement. It is an effective means to solve the heat environment measurement of hypersonic wind tunnels. It is one of the current development directions of hypersonic aerodynamic heat measurement. Phosphor Thermography system the working principle is as follows: the phosphor coating is sprayed on the surface of the model, and the model is heated by the hypersonic airflow during the wind tunnel test. An ultraviolet light source is used to illuminate the phosphorescent coating, and the coating radiates visible light at this time. The intensity of radiation depends on the intensity of ultraviolet light and the local surface temperature of the model. The surface temperature can be obtained according to the relationship between the calibrated radiation intensity and temperature, and the heat flow can be calculated.

Phosphorescent materials contain temperature-sensitive luminescent molecules, and their luminous quantum efficiency decreases with the increase of temperature. This temperature-related effect is thermal quenching. Within a certain temperature range, the relationship between light intensity I and absolute temperature can be described by Arrhenius equation:

$$\ln \frac{I(T)}{I\left(T_{ref}\right)} = \frac{E_{nr}}{R} \cdot \left(\frac{1}{T} - \frac{1}{T_{ref}}\right) \tag{7.1}$$

In this equation:

E_{nr} Activation Energy of Non-radiation Process;
R Universal Gas Constant;
T_{ref} Reference Temperature.

The selection and research of luminescent materials for wind tunnel phosphorescence thermography technology abroad are based on the luminescent characteristics and luminescent principles of rare earth ions at present. Rare earth elements have excellent spectral properties, and the luminescence of their ions is closely related to the surrounding environment (such as temperature). Relevant data of the ambient temperature can be obtained through monitoring and analyzing the changes of the optical characteristics of a certain luminescent material (such as intensity, wavelength, etc.), thus realizing quantitative testing of the external temperature.

Due to the intuitive luminescence of phosphorescent materials, diversified wavelengths and short reaction time (10^{-15}–10^{-9} s) etc., it has great advantages in studying the local or overall heating changes of high-speed moving objects and objects with complex geometric shapes under instantaneous and stable conditions, and has great application value in aerospace, national defense industry and civil fields. The phosphorescent materials adopted in this book all are rare earth elements as the luminescent component mainly.

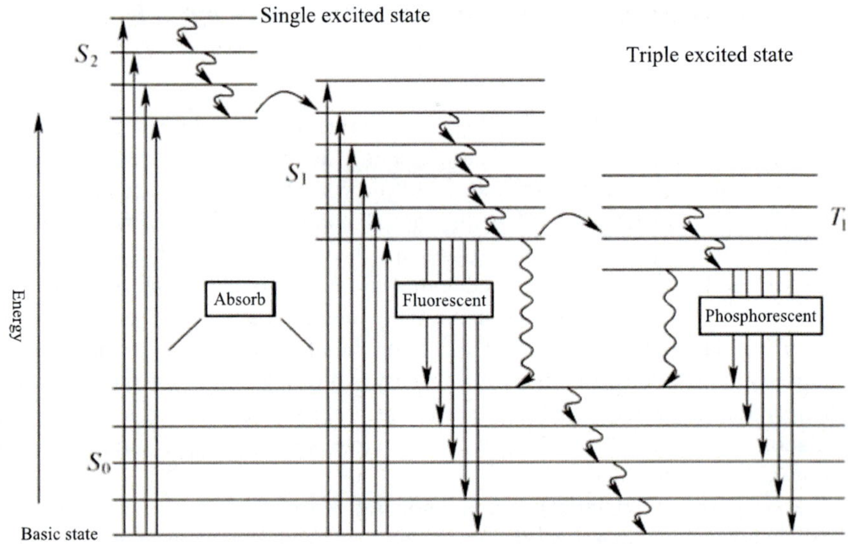

Fig. 7.4 Energy level spectrum of phosphorescent materials

The luminescence principle of phosphorescent materials is shown in Fig. 7.4. The atomic energy level of phosphorescent emission has two states of base state (*S0*) and excited state. Excited states include single excited state (*S1*, *S2*) and multiple excited state (*T1*). *S2* single excited state has the highest energy level as can be seen in the figure, S1 is in the second place, and triplet excited state T energy level is the lowest. The energy level of the atom can transition from the ground state to the excited state through the input of energy, which can be expressed as:

$$S_0 + hv \rightarrow S_1 \tag{7.2}$$

In the equation:

h Prandtl constant;
v The frequency of incident energy.

It may be accompanied by the process of radiating energy out when the electron energy level falls from the excited state to the ground state:

$$S_1 \rightarrow S_0 + hv_f \tag{7.3}$$

In the equation: *v*—radiation light frequency.

The light emitted is called fluorescence when the electron energy level changes from the single excited state S_1 to the ground state S_0, and the light emitted from the triplet excited state T_1 to the base state S_0 is called phosphorescence.

Certain other conversion processes that do not emit visible light are referred to as internal conversion (IC) or external conversion (EC), which are usually accompanied by a heat release process that does not emit visible light, such as by S1 To T1 Status:

$$S_1 \rightarrow T_1 + \Delta \tag{7.4}$$

In the equation: Δ—heat.

Externally converted high-energy particles interact with external environmental particles (such as solvent particles) to exchange energy, which causes the particles to fall to the ground state without radiating energy outward.

The intensity of visible light radiated outward by the phosphorescent material undergoes energy level transition changes with the change of the ambient temperature. The reason is that the frequency of collision between particles increases with the increase of temperature, which increases the probability of external conversion and reduces the number of particles that effectively emit light, thus reducing the luminous intensity.

The special luminescence properties of phosphorescent materials are used in temperature measurement experiments, that is, the temperature distribution of the environment where the phosphor is located is calculated according to the luminous intensity of the phosphor powder. The basic idea is shown in Fig. 7.5. The phosphor powder is sprayed on the surface of the model to form a uniform coating, and a certain wavelength of light is selected as the excitation light to irradiate the phosphorescent coating to continuously emit visible light. The model is heated by high-speed

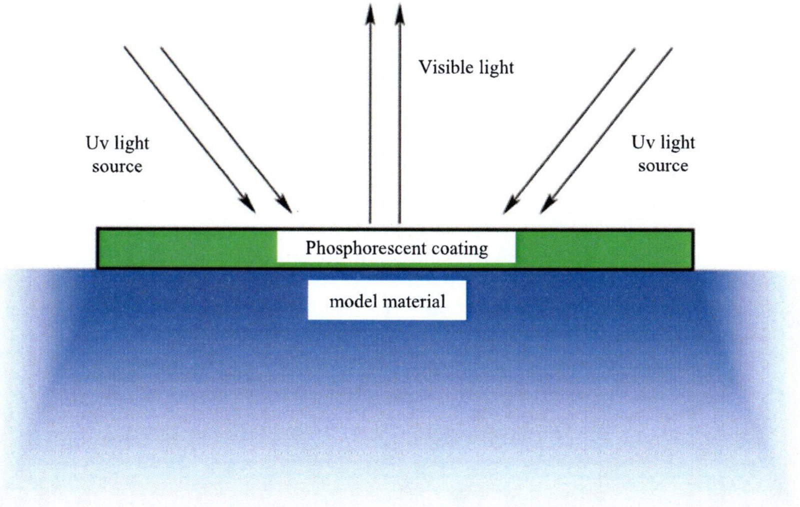

Fig. 7.5 Schematic diagram of phosphor thermography

airflow in the wind tunnel test, and different heat flows produce different temperatures the intensity of visible light radiated by the coating also changes accordingly. This change in light intensity can be recorded and stored by the image acquisition device. The radiation intensity is only related to the local surface temperature of the model when the intensity of the excitation light source is constant. According to the corresponding relationship between the calibrated radiation intensity and the temperature, the surface temperature distribution of the model can be obtained through the light intensity distribution, thus calculating the surface heat flow distribution (Figs. 7.6 and 7.7).

This thermal measurement technology is called monochromatic phosphorescence thermogram technology based on the relationship between intensity and temperature of monochromatic light. It is called monochromatic phosphorescence thermogram technology. Two-color phosphorescence thermogram technology is currently being developed abroad compared with monochrome phosphorescence thermogram technology. The difference between the two is that two-color phosphorescence powder

(a) (b)

Fig. 7.6 Side and front electron microscope photos. **a** Side electron microscope photo, **b** front electron microscope photo

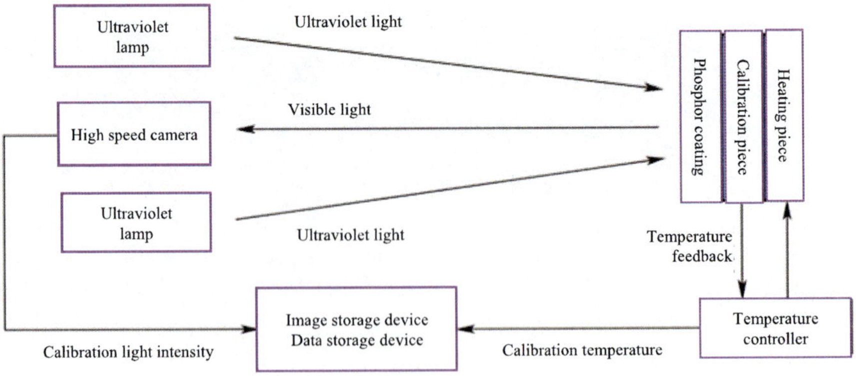

Fig. 7.7 The static calibration device and process of phosphor thermography

emits two different wavelengths of light after being irradiated by excitation light, and the thermal environment is measured according to the relationship between the light intensity ratio of the two colors of light and temperature. It can effectively reduce the system error caused by the uneven light intensity distribution and the acquisition error of the camera chip in the monochromatic phosphorescent system since the dual-color Phosphor Thermography Technology uses the light intensity ratio to measure temperature. The development of this technology requires more based on the mature monochromatic calorimetry technology, and it is realized through the research and development of new phosphorescent materials, adding cameras and adjusting the data transactions program.

7.2.3 Measurement Methods and Key Technologies of Phosphor Thermography Technology

7.2.3.1 Phosphorescent Coating Molding Technology

The surface of the model is scoured by high-speed airflow and bears great aerodynamic effects during the hypersonic wind tunnel experiment, and the phosphorescent luminescent material needs to be fixed on the surface of the model to go through the entire experimental process. Therefore, the phosphorescent luminescent material coating must be strong, erosion-resistant and be with other properties.

The phosphorescent coating is used as a temperature probe at the same time, and the heat resistance cannot be too large to affect the measurement accuracy of the heat flow on the surface of the model, so its thickness should also be strictly limited.

A liquid silicone resin material is determined as the curing medium of phosphorescent material on the surface of the model through in-depth research on the molding process of phosphorescent coating. This adhesive has low visible light absorption rate, high temperature resistance (300 °C) and strong thermal conductivity, and is completely suitable for being used as an adhesive in pulse wind tunnel. It was determined that the phosphor particles were uniformly distributed on the surface of the model by spray gun spraying, in order to obtain a uniform coating. The scale of phosphorescent materials, binders and thinners was continuously tested in the process of process research, and the spraying time, air pressure, spraying flow and other methods were adjusted to improve the various technical indicators of the coating, and finally a phosphorescent coating with excellent performance was obtained.

The phosphorescent coating currently used in the test is observed under an electron microscope (Fig. 7.12). Its thickness is thin (20 m), and the phosphorescent luminescent particles are evenly distributed and the thickness is uniform ($\pm$2 m). Wind tunnel blowing tests show that the phosphorescent coating is relatively strong and can be continuously blowing tests for many times without replacement, thus improving the experimental efficiency and reducing the experimental cost.

7.2.3.2 Phosphorescent Coating Calibration Technology

One of the key problems that phosphorescent can obtain quantitative results is how to obtain the corresponding relationship between temperature rise and light intensity change rate, which is directly applied to convert the light intensity distribution obtained during the experiment into temperature distribution. This work is usually completed before the start of the experiment. To obtain an accurate change relationship curve can make the heat environment measurement results more reliable.

The calibration process is to use the calibration system to simulate the experimental conditions inside the wind tunnel under static conditions, including model materials, lighting and phosphor coating thickness, etc., to obtain the light intensity-temperature versus response in advance.

The hardware equipment of the calibration system includes ceramic calibration sheet, temperature controller, temperature sensor, heating plate and support device (Fig. 7.13).

The temperature controller controls the surface temperature of the calibration piece coated with the temperature sensitive coating through the heating piece. The temperature sensor installed on the surface of the calibration piece is connected with the temperature controller to form a temperature control loop, which can maintain the constant temperature of the calibration piece. The temperature-light intensity relationship curve can finally be obtained by setting a series of light intensity sampling temperatures.

7.2.3.3 Phosphor Thermography Measurement System

The Phosphor Thermography measurement system is shown in Fig. 7.8. It is necessary to use the high-speed image Acquisition device to obtain enough effective images in such a short time since the experimental time of the shock tunnel is very short (20–40 ms). In addition, high sensitivity requirements are put forward for image acquisition equipment due to the weak radiation intensity of phosphorescent coating. Therefore, we must choose scientific-grade digital products.

Scientific-grade digital products are divided into two categories: CMOS and CCD, and their own characteristics are determined by their own different principles. CMOS has the characteristics of high speed, which can reach hundreds of thousands of frames per second, but the noise is high. CCD acquisition speed is relatively slow, and the highest sampling speed can only reach several hundred frames per second, but its quantum efficiency is high and its noise performance is good. The high-speed camera is connected to the workstation through the data Acquisition card. The light intensity distribution is stored in the workstation in real time during operation.

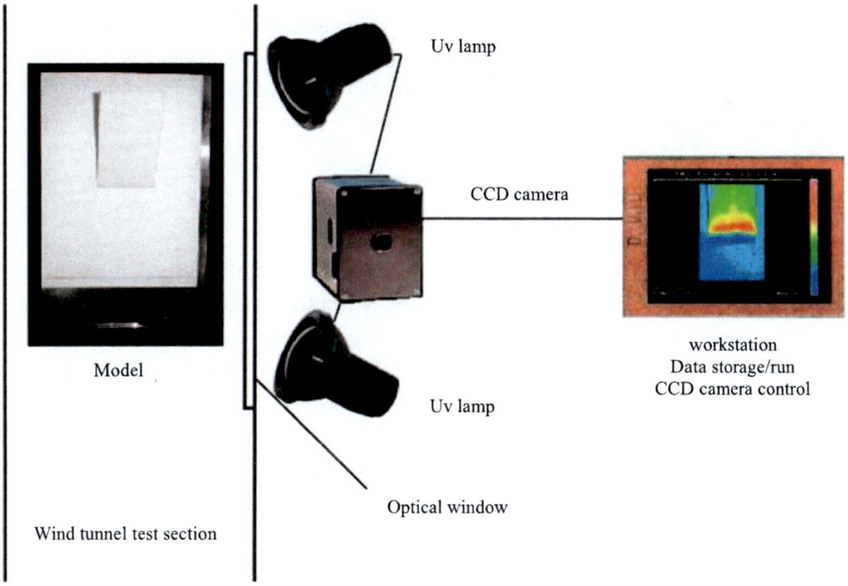

Fig. 7.8 Schematic diagram of phosphor thermography measurement system

7.2.3.4 Data Transactions Technology

The image is "contaminated" by the visible or invisible noise in varying degrees during the acquisition, encoding and transmission process.

Noise includes electronic noise, photon noise, speckle noise and quantization noise. The Noise will gradually become a visible grain contour if the signal-to-noise ratio (SNR) is lower than a certain level, which will cause the image quality to decrease, and filtering technology is required to remove most of the Noise points. The current Master filtering technology is divided into three types: spatial average filtering, median filtering and Wiener filtering. Each of the three filtering methods has its own characteristics. Median filtering method is a nonlinear spatial filtering method. Its operation is simple, easy to implement, and it can better protect the image boundary, compared with the other two filtering methods.

The conversion process from the light intensity distribution to the temperature distribution requires the use of a calibration curve (Fig. 7.9). The calculation process of temperature distribution for a certain image is as follows:

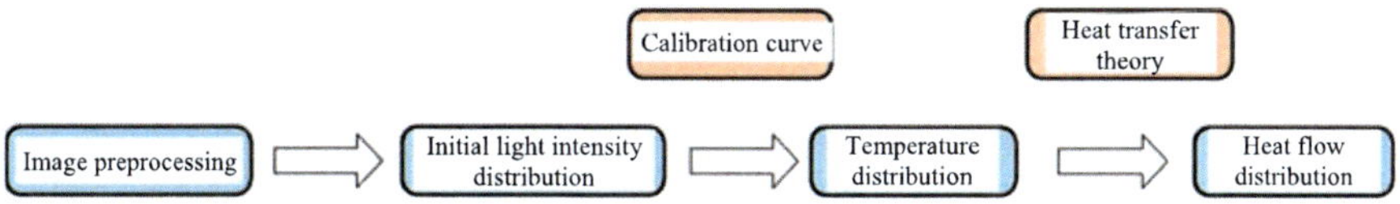

Fig. 7.9 Schematic diagram of phosphor thermography data transactions

First, the light intensity distribution of the flow initial frame is determined, and the light intensity distribution of a certain frame for temperature field calculation is determined *K* Then, the light intensity change rate distribution of frame *i* picture is obtained by using Eq. (7.5):

$$\frac{\Delta I}{I_0} = \frac{I_i - I_0}{I_0} \tag{7.5}$$

The temperature distribution of the corresponding pixel points can be calculated by bringing this light intensity change rate distribution into the calibration curve.

It is generally necessary to apply the one-dimensional semi-infinite medium assumption to solve the problem in the process of calculating the surface heat flow distribution from the model surface temperature rise distribution. The analysis model can be simplified to a two-layer dielectric heat transfer model considering that the model and the phosphor coating on the surface are two materials, as shown in Fig. 7.10.

There is a layer of medium 1 with a thickness of I on the semi-infinite medium 2, which is heated by the surface, and the heat flux is q_w. Medium 1 is a phosphorescent coating, and medium 2 is a model among them (Fig. 7.11).

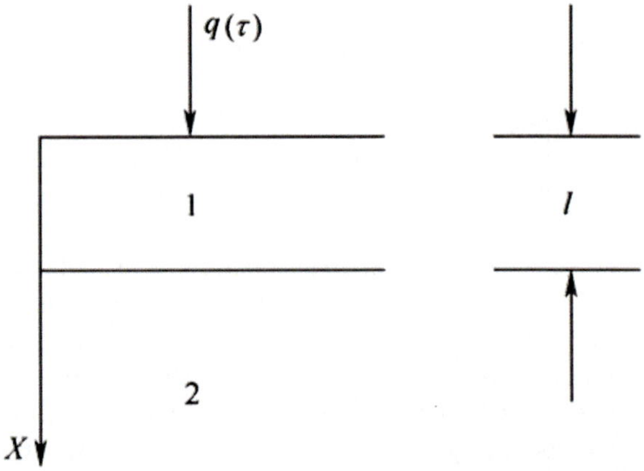

Fig. 7.10 One-dimensional semi-infinite medium heat conduction (double-layer structure)

Fig. 7.11 Compressed corner model

The Coerced equation for heat transfer in two-layer media is that:

$$\frac{\partial^2 T_1}{\partial^2 x} = \frac{1}{\alpha_1}\frac{\partial T_1}{\partial \tau} \tag{7.6}$$

$$\frac{\partial^2 T_2}{\partial^2 x} = \frac{1}{\alpha_2}\frac{\partial T_2}{\partial \tau} \tag{7.7}$$

In the equation: $\alpha = \frac{k}{\rho c}$, In which K, p, c are the heat conductivity, density and specific heat capacity of the medium respectively. Subscripts 1, 2 represent media 1, 2, respectively;

T The temperature rises from the surface x.

The boundary conditions are as the following If the surface heat flow rate is $q(t)$

At $x = 0$,

$$-k_1\frac{\partial T_1}{\partial x} = q_w \tag{7.8}$$

At $x = l$,

$$k_1\frac{\partial T_1}{\partial x} = k_2\frac{\partial T_2}{\partial x}, T1 = T2 \tag{7.9}$$

At $x = \infty$,

$$T_2 = 0 \tag{7.10}$$

After Laplace transformation of Eqs. (7.9) and (7.10), the general solutions are obtained:

$$\overline{T}_1 = \frac{\overline{q}\sqrt{\alpha_1}\left[(1+a)e^{-(x-l)\sqrt{P/\alpha_1}} + (1-a)e^{(x-l)\sqrt{P/\alpha_1}}\right]}{k_1\sqrt{P}\left[(1+a)e^{l\sqrt{P/\alpha_1}} - (1-a)e^{-l\sqrt{P/\alpha_1}}\right]} \tag{7.11}$$

$$\overline{T}_2 = \frac{2\overline{q}\sqrt{\alpha_1}e^{(l-x)\sqrt{P/\alpha_2}}}{k_1\sqrt{P}\left[(1+a)e^{l\sqrt{P/\alpha_1}} - (1-a)e^{-l\sqrt{P/\alpha_1}}\right]} \tag{7.12}$$

In the equation: $a = \sqrt{\frac{\rho_2 c_2 k_2}{\rho_1 c_1 k_1}}$

The ratio of its thickness to the model thickness is $l/L \to 0$, and its ρck value is much larger than that of the second layer considering that the medium 1 is very thin.

It is approximately considered the 1 close to 0 in the case of large (i.e. $a \to 0$), then the Eq. (7.12) can be simplified as:

$$\bar{T} = \frac{1}{\sqrt{\rho ck}}\frac{\bar{q}}{\sqrt{P}} \tag{7.13}$$

Equation (7.13) is transformed through the inverse Laplace, we can get:

$$q(\tau) = \frac{\sqrt{\rho c k}}{\pi} \int_0^t \frac{dT(\tau)/d\tau}{\sqrt{t-\tau}} d\tau \tag{7.14}$$

The relationship between surface temperature T and time t is recorded in the experiment, converted into the relationship between surface temperature change rate dT(t)/dt and time t, and then integrated in the time domain [0, t]. The surface heat flow in the model at time t can be obtained.

But in the actual experiment, T is recorded instead of $dT(\tau)/d\tau$. Therefore, z:
Let $z = T(t) - T(\tau)$, then $\frac{\partial z}{\partial \tau} = -\frac{\partial T(\tau)}{\partial \tau}$.
The following can be obtained by simple transformation,
When $\tau = t, z = 0$; when $\tau = 0, z = T(t)$, then Eq. (7.15) can be written as:

$$(\tau) = \frac{\sqrt{\rho c k}}{\pi} \int_0^t -\left\{ \frac{d}{d\tau}\left[\frac{z}{\sqrt{t-\tau}} \right] - \frac{1}{2}\frac{z}{(t-\tau)^{3/2}} \right\} d\tau \tag{7.15}$$

Equation (7.16) is the heat flow calculation equation of phosphorescence thermography technology, where the amount of temperature change at the beginning of the time $t = 0$ for $T(t)$. The temperature rise is calculated, and then the temperature value at the initial time of flow is subtracted to obtain the temperature $T(T)$ according to the calibration curve. The integration method is beneficial to reduce the interference of individual image errors on the overall results.

The duration of the hypersonic gun wind tunnel experiment is only tens of milliseconds. The temperature gradient along the tangent to the surface is much smaller than the normal temperature gradient perpendicular to the surface. Therefore, it is approximately considered that the conduction of heat flow can be regarded as one-dimensional. Apply the heat flow calculation equation derived from the heat transfer equation of one-dimensional semi-infinite body. There are two methods to solve Eq. (7.16): Cook-Felderman method and Kendall-Dixon method. The following two algorithms are introduced and compared respectively.

(1) Cook-Felderman method

This heat flow algorithm is currently a relatively common algorithm in the world, and it is also the method used by the current data transactions. The temperature can be regarded as a function of time under the assumption that the heat transfer property of the coating material is constant. The relationship between surface temperature $T(t)$ and heat flow $q(t)$ is finally obtained through Laplace transformation and inverse transformation of the one-dimensional heat conduction equation, as shown in Eq. (7.16).

The surface temperature is calculated from the discrete single picture at each time in the experiment, so the discrete scheme that can be applied to the experiment is obtained after partial linearization of the function $T(t)$ in Eq. (7.16):

$$q(t) = \frac{\sqrt{k\rho c}}{\sqrt{\pi}} + \frac{1}{2}\int_0^t \frac{T(t) - T(\tau)}{(t-\tau)^{3/2}} d\tau \tag{7.16}$$

$$q(t) = 2\sqrt{\frac{\rho c k}{\pi}} \sum_{i=1}^{m} \frac{T(i) - T(i-1)}{(t_m - t_i)^{1/2} + (t_m - t_{i-1})^{1/2}} \tag{7.17}$$

(2) Kendall-Dixon method

This method is partly based on Cook-Felderman method and is divided into two steps. First, the time is calculated with integral, so as to obtain the cumulative amount of energy entering the system during the experiment:

The second step is to apply the high-order central difference scheme to obtain the heat flow at a certain moment after discretization:

$$q'(t) = \sqrt{\frac{\rho c k}{\pi}} \sum_{i=1}^{m} \frac{T(i) + T(i-1)}{(t_m - t_i)^{1/2} + (t_m - t_{i-1})^{1/2}} \tag{7.18}$$

$$q(t) = \frac{-2q'_{t-8} - q'_{t-4} + q'_{t+4} + 2q'_{t+8}}{40\Delta t} \tag{7.19}$$

7.2.4 Analysis on Typical Wind Tunnel Test Cases of Phosphor Thermography Technology

The phosphorescence thermal diagram technology is used to analyze the heat transfer distribution law according to the compression corner model (Fig. 7.17) with three different wedge angles (20, 30 and 40), and repeated tests are carried out to obtain the repeatability test accuracy (Table 7.1). The measurement accuracy of phosphorescence thermography technology is compared and verified at the same time, combined with sensor technology and schlieren technology.

The repeatability experimental results of 20 wedge angle three-wheel phosphorescence thermography technology under the same incoming flow state (Fig. 7.12) show that the peak heat flow position of the center line and the heat flow level at the same position are basically the same. The calculation results of the data error show that the repeatability error of the heat flow results of the three Phosphor Thermography experiments with a 20° wedge angle is less than 5%, and the results of

Table 7.1 Compression corner test state

M_∞	Re_∞/1/m	P_0/MPa	T_0/K	Wedge angle/°
8.04	2.21×10^7	18.18	1066	20/30/40

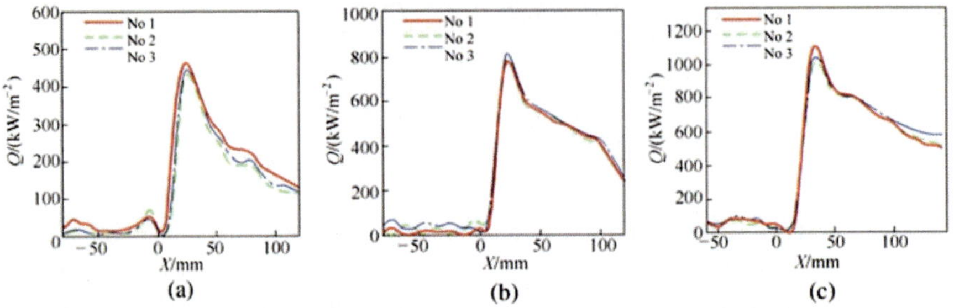

Fig. 7.12 Test results of heat flow repeatability of compression corner centerline at different angles. **a** 20° wedge angle; **b** 30° wedge angle; **c** 40° wedge angle

30° and 40° are basically consistent with the results of the 20° wedge angle, and the repeatability error is less than 5%.

The cloud images with different wedge angles show that with the increase of wedge angle, the peak heat flow in the reattachment area of the wedge surface increases and the area of the high heat flow area increases. The comparison results between each wedge angle and the sensor (the error is less than 15%) also show that the measurement results of phosphorescence thermography technology have high reliability.

Notably, in Fig. 7.13d, the measurement of the position and size of the peak heat flow on the wedge surface is not ideal in the comparative experiment results using heat flow sensor, due to the discrete distribution of sensor measuring points, while the continuous heat flow distribution curve obtained by phosphorescence thermography technology accurately captures the position and size of the peak heat flow on the wedge surface and shows the approximate shape and area of the high heat flow region.

The schlieren experiment shows that the wedge surface high heat flow area is the result of shock wave/boundary layer interference and separation flow reattachment, and the Phosphor Thermography Technology visually shows a certain internal connection between aerodynamic heating and flow structure through the heat flow distribution, as shown in Fig. 7.14.

Phosphorescence thermography technology can also be used for the research and analysis of flow structure as a surface measurement technology, and the analysis of flow field in interference area from thermal aspects is rarely reflected in previous studies, which is also an effective supplement to the traditional flow display technology of phosphorescence thermography technology "heat flow field".

The plate-protrusion is selected as the phosphorescence thermograph technology test model, and the flow structure in the interference area of the surrounding plate is changed by changing the shape of the protrusion, so that the heat transfer distribution on the surface of the model changes accordingly. The "heat flow field" is displayed and analyzed by phosphorescence thermograph technology (Table 7.2).

The test results (Figs. 7.15 and 7.16) clearly show the structure and law of the "heat flow field" around the protrusions.

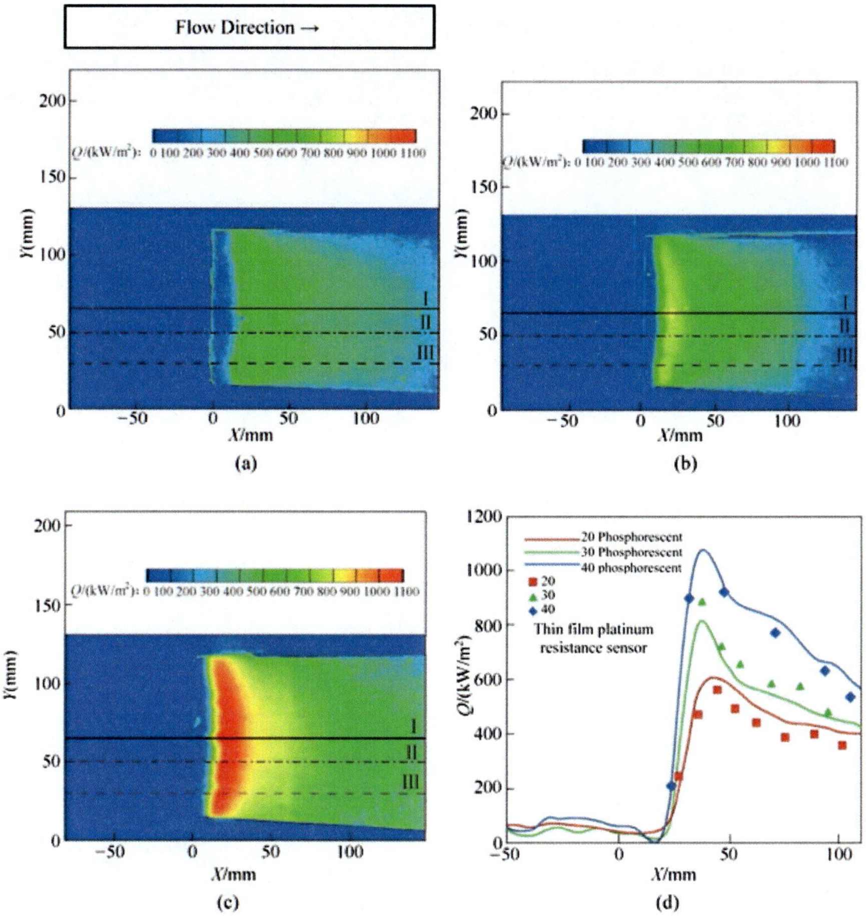

Fig. 7.13 Heat flow cloud diagram of compressed corners at different angles and its comparison with sensor technology. **a** 20° wedge angle cloud image; **b** 30° wedge angle cloud image; **c** 40° wedge angle cloud image; **d** comparison of ratio of centerline sensors with different angles

Taking the flat-cylinder shape as an example, the heat flow distribution cloud picture of the flat-cylinder shape (Fig. 7.14) is selected, and a clearer understanding of the flow structure in the interference area can be obtained by combining the cylinder flow theory and the oil flow display results (Fig. 7.15):

The flow around the front of the cylinder is a typical double vortex structure. There is a saddle-contoured separation line S induced by the separation shock in the far front of the cylinder, and a bow-contoured shock wave B is generated downstream of the separation line, in the heat flow field cloud image It can be seen that an area with higher heat flow is formed around this line than the front. The heat flow peak appears on the reattach line A position with the reattachment of the separation flow, forming a high heat flow area and extending to the rear of the cylinder.

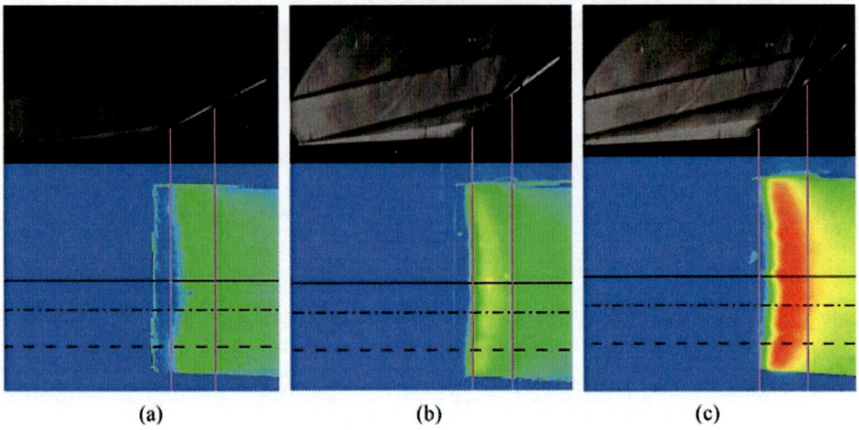

Fig. 7.14 Comparison of heat flow cloud picture and schlieren results of compressed corners at different angles. **a** 20° wedge angle; **b** 30° wedge angle; **c** 40° wedge angle

Table 7.2 Test state of flat plate-protrusion

M_∞	Re_∞/1/m	P_0/MPa	T_0/K
8	2.21×10^7	18.2	1077

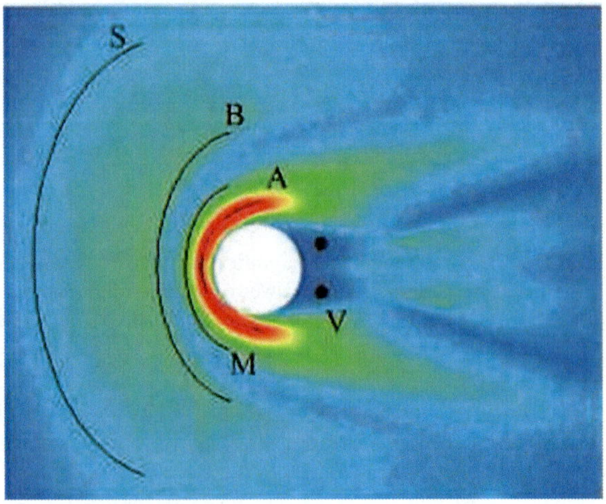

Fig. 7.15 Display and analysis of plate-cylinder thermal flow field (Feature Dimension 30 mm) heat flow cloud atlas

However, there is a relatively obvious low heat flow region around the "vortex core" V behind the cylinder, while the downstream is affected by the vortex core compression effect, forming two symmetrical compression waves, increasing the heat flow in the wake region and forming a "dovetail" shape wake structure.

The "heat flow field" of the flat-square column and the flat-diamond-contoured column also shows the changes in the flow structure of the interference zone caused

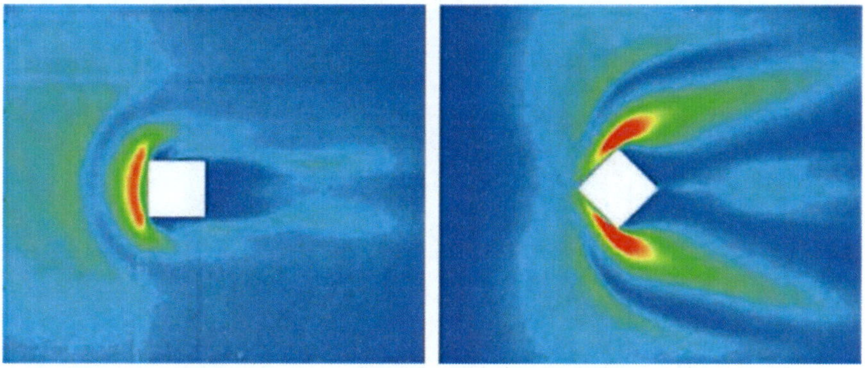

Fig. 7.16 Heat flow field display and analysis. **a** Plate-square column heat flow cloud atlas; **b** plate-diamond column heat flow cloud atlas

by the characteristics of their respective contours (Fig. 7.16). We will not do too much analysis here, but it can be seen that the role of the Phosphor Thermography Technology is not only limited to the size of the heat flow on the surface of the model and its distribution, its surface measurement characteristics make it can be used as a flow display method to analyze the flow field structure of the model surface.

The ultimate purpose of the ground wind tunnel thermal environment test is to serve the thermal protection design of the HAVs. It is also the main development direction of phosphorescence thermography technology to provide comprehensive and effective test data for it. The full flow field thermal environment test for complex shape aircrafts is also the best way and the ultimate goal to test its comprehensive test capability. Based on this, taking the X-33 shape aircraft as the blueprint and comprehensively applying the processing technology of the complex shape ceramic model, the processing of the ceramic model imitating the X-33 shape aircraft was completed (Fig. 7.17). The phosphorescence thermogram test verification research was carried out according to its different flight states (angle of attack: 0/4). The test states are shown in Table 7.3.

Phosphorescence thermography technology was applied to measure the heat flow distribution in the back region, lateral region and airfoil/control surface of the model, and the overall heat flow distribution and the detailed heat transfer characteristics of the surface of local complex components were observed during the experiment.

The results of heat flow measurements on the upper and lateral surfaces of the model at 0° angle of attack show that, as can be seen from Fig. 7.18a, there are obvious heating conditions in the root of the oblique wing, the leading edge of the rudder, the head of the model and both sides of the model. This is also confirmed in the cloud picture of the side heat flow distribution in Fig. 7.18b, and a high heat flow area with slightly higher heat flow than the surrounding fuselage can be observed in the fuselage/oblique wing interference area in Fig. 7.18b, extending along the flow direction below the oblique wing to the tail of the model.

Fig. 7.17 Imitation X-33 ceramic model

Table 7.3 Configuration test status of imitation X-33

M_∞	Re_∞/(1/m)	P_0/MPa	T_0/K	AoA/(°)
8	5.2×10^6	5.1	1183	−4/0/4

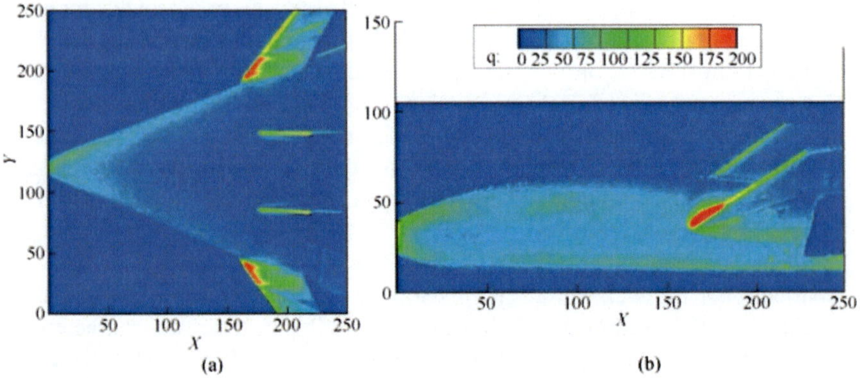

Fig. 7.18 Test results of phosphorescence thermogram at different positions at 0° angle of attack. **a** 0° forward; **b** 0° lateral

The phenomenon worthy of attention is the complex heat flow field on the inclined wing surface. As can be seen from Fig. 7.18a, there is an obvious high heat flow line extending from the leading edge of the wing to the trailing edge of the wing, penetrating the entire wing surface and independent of the high heat flow area at the wing root. The position with this high heat streamline coincides with the position

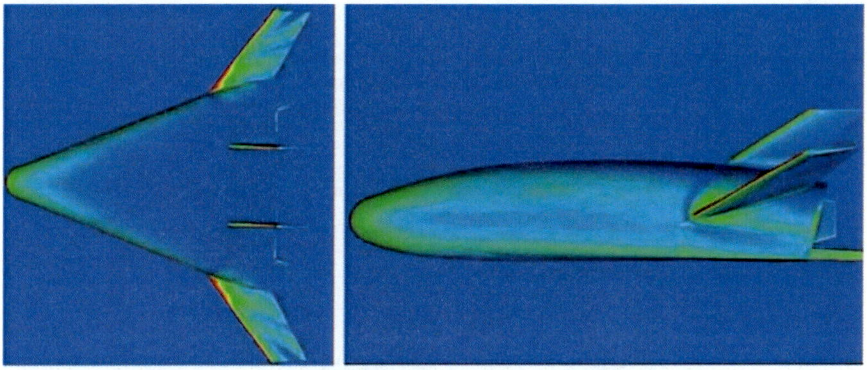

Fig. 7.19 Numerical simulation results of different positions at 0° angle of attack. **a** 0° forward; **b** 0° lateral

(Fig. 7.19) Height where the head shock wave interacts with the wing leading edge shock wave in the NASAX-33 schlieren test.

It can be seen that the results of the two are in good agreement, and the heat flow distribution laws of complex interference areas on the airfoil and rudder surfaces are consistent through the comparison with the numerical simulation technology (Fig. 7.20), which reflects the accuracy and reliability of phosphorescence thermography technology from the side.

The phosphorescence results at an angle of attack of ±4° (Fig. 7.21) show the change of the heat flow field on the upper surface of the model and the airfoil with different angles of attack. The heat flow of the inclined wing surface, the leading edge of the rudder and the leading edge of the model, especially in the root area of the wing surface, increases significantly as the angle of attack decreases, but the high heat flow area position and the contour have not changed significantly, indicating that the amount of pneumatic heating in this state is different, but the law is the same for the model is in these two areas.

7.2.5 Summary and Prospect of Phosphor Thermography Technology

The non-contact measurement characteristics of the Phosphor Thermography Technology can be effectively applied in the heat environment experiment of the complex contour aircraft. The reasonable application of the temperature measurement ability of the phosphorescent material (theoretical up to 1700Y) can extend this technology to a larger range aircraft heat environment test research and hypersonic aircraft high heat flux area measurement work (such as the field of heat environment measurement under high temperature in arc wind tunnel).

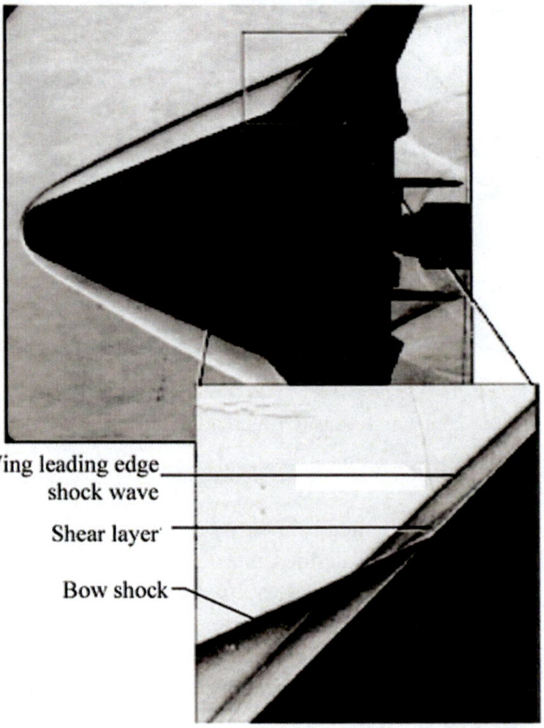

Fig. 7.20 X-33 Schlieren test results

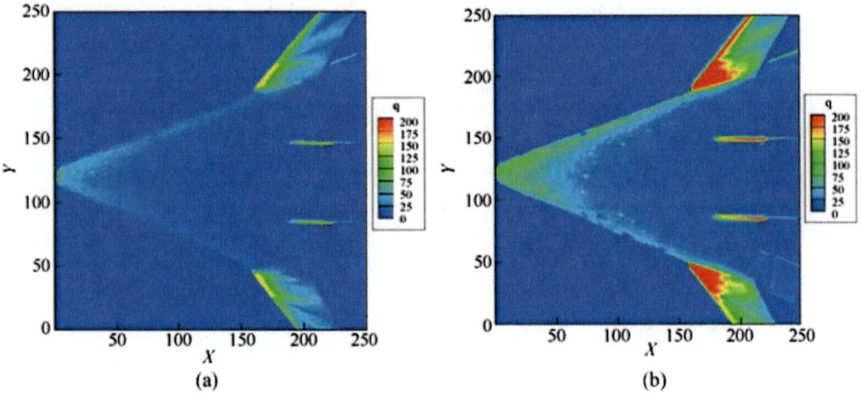

Fig. 7.21 phosphor thermography test results at different angles of attack. **a** +4° angle of attack (head up); **b** −4° angle of attack (head down)

Wind tunnel heat flow measurement is required to provide higher precision and more comprehensive heat flow data with the rapid development of domestic aerospace technology, and the technical advantages of Phosphor Thermography Technology will inevitably become an indispensable experiment for future ground heat measurement experiments. Technology is also the general direction of development in the field of heat flow measurement in the future.

7.3 Particle Image Velocimetry

Fluid flow is a very complex natural phenomenon. We can know the flow characteristics of the flow field by observing fluid flow, which plays an important role in the research and development of fluid mechanics. The study of fluid forces and motion laws in fluid mechanics needs to determine the relationship between the various flow parameters and the mutual values in the flow process, which is inseparable from the measurement of flow parameters. Flow display technology can visualize the flow process, and flow measurement technology can obtain quantitative information of fluid transmission through some method. Both of them constitute the main content of experimental fluid mechanics and continuously serve the development of theory and the improvement of experimental technology.

Experimental fluid mechanics puts forward higher requirements for flow display and measurement technology with the development of fluid mechanics. Smoke line method and dyeing solution method can intuitively describe the flow image of the flow field although traditional flow display methods such as hydrogen bubble method, they cannot obtain the flow parameter information of the flow field. Velocity measurement methods such as Laser Doppler Velocimetry (LDV) and Hot-Wire Anemometry (HWA) have been developed on the basis of traditional measurement methods (Pitot tube). They have high time resolution and have provided important measurement technologies for turbulence research. However, LDV and HWA are single-point measurement technologies, which obviously cannot meet the needs of complex flow conditions and full-field rate measurement, and the intervention of external media (hot wires, probes, etc.) disturbs the local flow field [20].

Particle Image Velocimetry (PIV) is a non-contact full-field measurement technology. It began to develop in the 1980s, and its technical basis came from Laser Speckle Velocimetry (LSV) in the mid-1970s. The predecessor of Laser Speckle Velocimetry is Laser Speckle Interferometry, which is a measurement technology used to detect the deformation of composite materials through optical interference in engineering structure research [21]. The technical principle of LSV is to spread high concentration tracer particles in the flow field, illuminate the flow field with a sheet light source, and record the speckle image of tracer particle swarm on the negative for two consecutive exposures. Optical interference analysis is carried out on the speckle image of tracer particle swarm to obtain the displacement information of the speckle image. LSV has derived two technologies with the change of the concentration of tracer particles, one is Particle Tracing Velocimetry (PTV), which is suitable for the situation where the concentration of tracer particles is very low. The method of particle identification and tracking is to determine the rate of the tracer particles one by one using a single tracer; the other is PIV, which is suitable for the situation where the concentration of the tracer particles is medium, and calculates the statistical average rate of all tracers in the minimum resolution volume method. It can be seen that the difference between LSV, PIV and PTV lies in the concentration of tracer particles and the corresponding rate transactions method [22].

The development of early PIV technology has certain limitations. Particle images recorded on optical film need to be processed by dark room and optical interference to determine the displacement in the shooting flow field, which will increase the errors contained in the calculation results, lead to long processing period and poor efficiency, and can only be used for studying properties. The sign of PIV technology becoming practical is the application of digital recording and digital transactions technology in PIV. The intervention of digital camera (CCD) makes PIV technology gradually deviate from the traditional optical film recording method, combined with the computer's digital transactions technology, greatly shorten the period of image recording to transactions, and improved the experimental accuracy, spatial and temporal resolution and experimental efficiency.

In particular, the emergence of digital frame-straddle cameras greatly increases the range of rate that can be measured by PIV. The cross-frame interval as low as several hundred nanoseconds makes it possible to measure rate in supersonic and even hypersonic flow fields.

It will bring great projection error to the two-dimensional measurement results when the vertical velocity component is large, since the two-dimensional particle image velocimetry technology only obtains the projection result of the three-dimensional velocity vector in the light plane of the measured piece. Stereo Particle Image Velocimetry (Stereo Particle Image Velocimetry, SPIV) can fundamentally avoid the errors caused by the above perspective projection, and more importantly, the velocity component perpendicular to the measurement plane can be obtained through the synthesis of the data of the two cameras, which is very useful for the measurement of complex flow fields with obvious three-dimensional flow effects. The SPIV system is different from the ordinary two-dimensional PIV system, SPIV uses two cameras to photograph the plane to be measured in the flow field from two off-axis directions, record particle images, and then obtain the three-component velocity field of the tangent through corresponding image processing and velocity reconstruction after dispersing a certain concentration of tracer particles in the flow field. Its typical layout is divided into translation layout and angular shift layout according to the included angle between the lens and the plane to be measured.

The two structural layouts have their own characteristics. The advantage of translation layout is that layout is simple and easy to implement, ordinary camera can be competent for shooting without structural adjustment, and the overlapping area of the two cameras can be reconstructed and calculated by using a unified magnification in the result processing since the two cameras are equidistant and parallel to the shooting plane [23–25]. However, the translational layout also has corresponding shortcomings: firstly, the imaging overlap area of the two cameras is small, resulting in a larger area of images being wasted under normal shooting conditions; secondly, the out-of-plane rate component accuracy cannot be further improved due to the limited field of view imaging angle of the lens.

The angular displacement layout mechanism is based on the premise of satisfying Scheimpflug condition, which avoids the deficiency of the translation layout mechanism structurally. The Scheimpflug condition is stated as: the object plane, lens plane and image plane or their extension lines to be photographed must intersect at one point, so as to ensure that all points on the object plane can be clearly imaged on the image plane, but image correction and rate are required. Reconstruction work [26–28].

Modern digital technology endows PIV with stronger adaptability and broader development space. PIV has expanded from two-dimensional velocity measurement to three-dimensional velocity measurement, and has been widely used in the fields of fluid mechanics, medicine, chemistry, ships and mechanical manufacturing.

7.3.1 Principle and Method of Particle Image Velocimetry

The current two-dimensional particle image velocimetry technology has been very mature. Its basic principle is: uniformly spread tracer particles of a specific density in the measured flow field, so that each minimum resolution volume contains 4–10 particles, light illumination flow field with pulsed lasers manufactured sheet, use optical film or digital camera to shoot and record the illumination flow field test segment image through two or more exposures, use optical Young's fringe method or cross-correlation method in image transactions to calculate each interpretation, the two-dimensional rate of the tangential flow field is calculated according to the laser exposure time interval according to the statistical average displacement of the particles in the area.

$$V = \frac{\Delta x}{\Delta t} = \frac{x(t_1) - x(t_2)}{t_1 - t_2} \tag{7.20}$$

In this equation:

$x(t_1)$ the particle space position(m) at t_1;
$x(t_2)$ the particle space position(m) at time t_2;
$t_1 - t_2$ Corresponding time interval (s).

Traditional optical interference analysis methods have gradually withdrawn from the stage of PIV applications with the emergence of digital cameras, especially cross-frame technology. Digital recording technology has the characteristics of fast recording rate, high measurement accuracy, and easy data analysis transactions. The digital camera using cross-frame technology makes it possible to measure supersonic and even hypersonic flow fields. The statistical average displacement of the particle image in a specific area is obtained by image correlation technology in the digital particle image velocimetry (DPIV) technology, and the time interval is determined according to the exposure interval of the two pulses of the laser, and finally the rate of the flow field is obtained.

Correlation technologies of image processing include autocorrelation technology and cross-correlation technology. They adopt different calculation methods according to different image recording modes: Autocorrelation technique is applied to the case where two or more exposed images are recorded on the same negative. The interpretation area of the image searches for the most relevant area in an image and determines the statistical displacement. However, the autocorrelation method has the ambiguity of velocity direction. The velocity measurement range is very small although it can be solved later according to the flow characteristics or by applying the displacement offset method. Cross-correlation technology is applied to the case where an image with two exposures is recorded on two negatives. Among them, the determined interpretation cell in one frame image searches the region with the greatest correlation with it in the other frame image to determine the statistical displacement of particles, thus improving the signal-to-noise ratio and automatically distinguishing the velocity direction. The range of velocity measurement is wider than that of autocorrelation method. Fourier transform is a basic tool for modern signal and image processing. It converts image signals from time domain to frequency domain for correlation processing. However, the continuous improvement of computer performance and the development of fast Fourier transform (FFT) have greatly improved the computational efficiency.

The cross-correlation technology is applied to the situation where the images of the two exposures are recorded on two negative films, in which the determined area in one frame of image searches for the area with the greatest correlation with it in the other frame of image. The particle images exposed twice are recorded in two frames of images respectively at an interval of At in actual shooting. First, a fixed square area is given in the first image as an Interrogation area, and then the similarity degree of the two images is studied in the corresponding search area in the second image. It is an important process to distinguish the similar parts of the two images (using the gray value as the discriminant value) in the image transactions, which is called image matching. The statistical average displacement of particle microclusters can be obtained through the interpretation and cross-correlation calculation of the corresponding areas of the two images, and then the statistical average velocity of the measured flow field in the interpretation area can be obtained according to the exposure interval Δt.

Figure 7.22 illustrates the main process of cross-correlation image matching, wherein $1\ (x,\ y)$ and $r\ (x,\ y)$ represent the gray values of the points $(x,\ y)$ in the corresponding interpretation area and the search area, respectively. The definition of one-dimensional correlation function is given in digital signal processing, and the correlation calculation method originates from the time autocorrelation function in signal and linear system analysis. See the first row in Table 7.4. Its corresponding one-dimensional discrete and two-dimensional discrete forms have also been developed for correlation calculation of various digital signals.

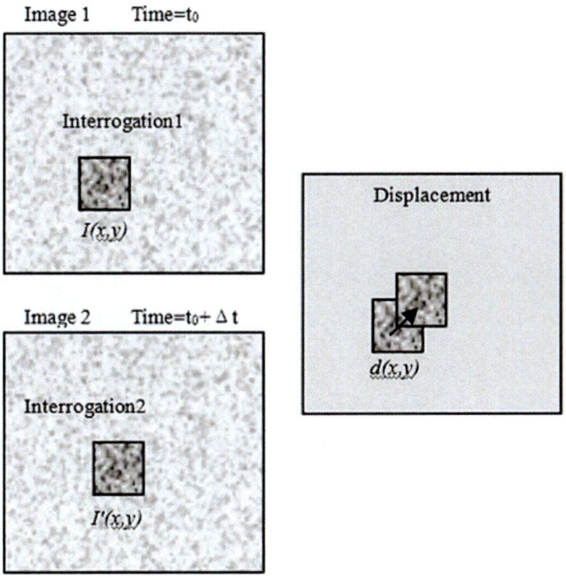

Fig. 7.22 Frame cross-correlation matching

Table 7.4 Signal characteristics and correlation function forms

Signal	Signal features	Correlation function
x(t)	One-dimensional continuous	$\overline{R(\tau)} = \int_{-\infty}^{\infty} x(t)x(t+\tau)dt$
$x[m]$	One-dimensional discretization	$\overline{R(\delta)} = \sum_{m=-\infty}^{\infty} x[m]x[\delta+m]$
$x[m]$ $y[m]$	Binary one-dimensional discretization	$\overline{R_{xy}(\delta)} = \sum_{m=-\infty}^{\infty} x[m]y[\delta+m]$
$x[m, n]$ $y[m, n]$	Two-dimensional discretization	$\overline{R_{xy}(\delta,\xi)} = \sum_{m=-\infty}^{\infty}\sum_{n=-\infty}^{\infty} x[m,n]y[\delta+m,\xi+n]$

The two-dimensional PIV image transactions is concerned with the two-dimensional discrete form of the cross-correlation function since the image pixels are two-dimensional distribution.

The image interpretation method shown in Fig. 7.23 combined with the given two-dimensional discrete signal correlation function can obtain the two-dimensional correlation function form based on image gray scale discrimination:

$$R_{II}(i,j) = I[i,j] \circ I'[i,j] = \sum_{m=0}^{M-1}\sum_{n=0}^{N-1} I\big[m,n\big]I'\big[i+m, j+n\big] \tag{7.21}$$

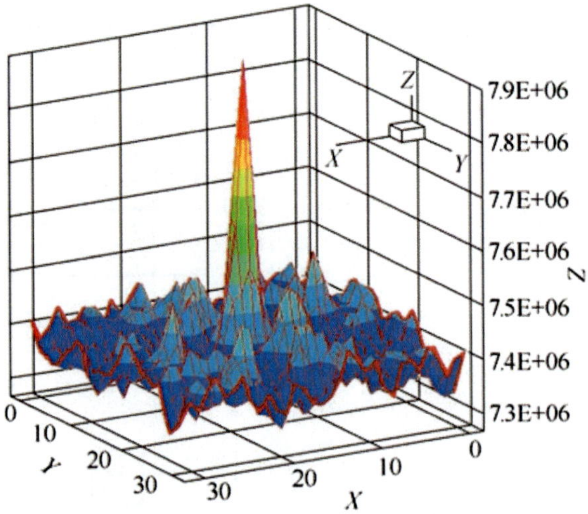

Fig. 7.23 Distribution of cross-correlation function values

In the equation:

M the width of the interpretation area;
N Interpretation area height.

Both M and N are in pixels. General use $\hat{r}_{I,I'}(i,\ j)$ to represent positive $R_{II}(i, j)$ for normalization, which is called correlation coefficient:

$$\hat{r}_{I,I'}(i,\ j) = \frac{\sum_{m=0}^{M-1}\sum_{n=0}^{N-1} I[m,n]I'[i+m,\ j+n]}{\sqrt{\sum_{m=0}^{M-1}\sum_{n=0}^{N-1} I[m,n]^2}\sqrt{\sum_{m=0}^{M-1}\sum_{n=0}^{N-1} I'[i+m,\ j+n]^2}} \tag{7.22}$$

The correlation function is the autocorrelation function when $I[i, j]$ and $I'[i, j]$ are the same signal, and the correlation coefficient is called the autocorrelation coefficient. R_{II}(i, j) and $\hat{r}_{I,I'}(i,\ j)$ are respectively called cross-correlation function and cross-correlation coefficient if the interpretation area and the search area are not the same image.

The cross-correlation function value distribution is shown in Fig. 7.23, in general, there will be extreme points in the range of the independent variables (x, y). You can further analyze the data of $R_{II}(i, j)$ in the correlation function to find the extreme value of position, and then the statistical average displacement of the particle clusters in the interpretation area of the two images before and after.

Figure 7.24 shows the correlation function value distribution of the aforementioned autocorrelation algorithm. It is necessary to find the second largest peak to be the displacement of the particle image of the two exposures in the calculation of the PIV rate field, since the autocorrelation calculation function is symmetry. There will be two sub-large peaks in the autocorrelation function distribution graph, causing rate directional ambiguity, and the cross-correlation calculation results, there is no

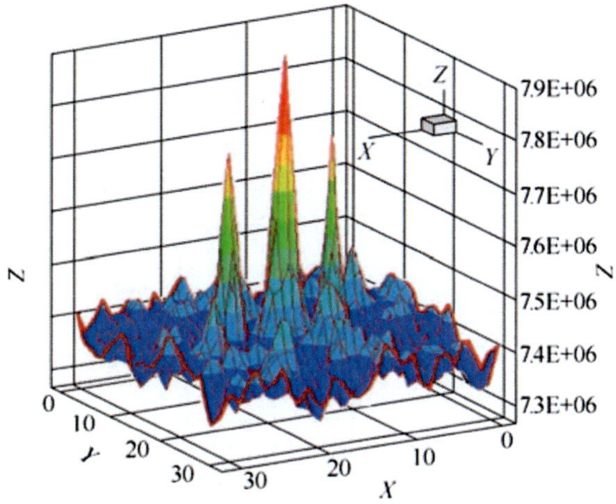

Fig. 7.24 Autocorrelation function value distribution graph

directional ambiguity problem, you can directly find that the extremely large peak position is the displacement of the particle image value.

The size of the interpretation area and the search area are set on a basis in the above-mentioned cross-correlation calculation method. The search area of the second image cannot be too small after the interpretation area of the first image is determined. It is difficult to calculate the correlation Get the displacement result if the search area of the second image is too small at a certain flow rate and exposure interval and it does not contain or contains few particle images that are strongly correlated with the interpretation area of the first image. The common practice is to have a certain estimation of the speed, and then guide the setting of the search area size of the second image according to the displacement of the speed within a fixed exposure interval. Generally speaking, it is better to estimate the displacement value less than 1/3 of the interpretation area.

However, this kind of estimation is sometimes unreliable, because sometimes the estimated rate value is a range and has no guiding significance, and different areas may have different search area sizes. The general approach is to enlarge the search area of the second image, for example, the size is 2–3 times of the interpretation area of the first image, and then move the first image interpretation area in the large search area of the second image in turn to identify the particle image that best matches the first image interpretation area (Fig. 7.25). In this way, the cross-correlation function distribution diagram can still be obtained, and the displacement of the two particle images before and after can be obtained by judging the position with the maximum correlation value.

Obviously, this method of expanding spatial domain search effectively improves the calculation accuracy and resolution of velocity vector, especially for the flow field with large velocity gradient, and because the size of the first interpretation cell is relatively reduced and the proportion of effective particle pairs is increased, good signal-to-noise ratio can be obtained.

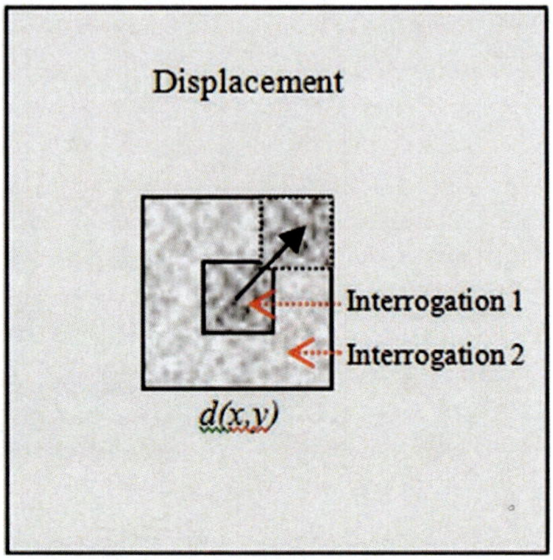

Fig. 7.25 Extended search area interpretation method

DPIV cross-correlation calculations can also be calculated using digital signal analysis tools in the complex frequency domain in addition to direct correlation calculations in the space domain. The most commonly used one is the fast Fourier transform technology (Fig. 7.26). As can be seen from above related calculation method, in condition that the interpretation area (M X N) size is fixed, when the size of the interpretation area and the search area are the same, the calculation amount of cross-correlation and autocorrelation is $(MN)^2$. The number of multiplication operations is very large. When the interpretation area is smaller than the search area ($L \times H$) and the calculation amount of cross-correlation and autocorrelation is $(L-M+1)(H-N+1)MN+LH$ multiplication operation (since the size of the correlation coefficient distribution matrix should be the same regardless of whether the size of the interpretation area is less than or equal to the size of the search area, so $L-M+1=$

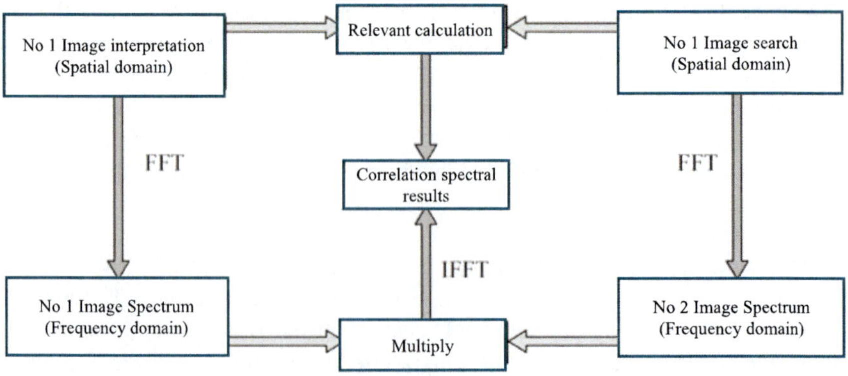

Fig. 7.26 Schematic diagram of FFT algorithm for DPIV

M, H-N + 1 = *N),* the amount of computation is even larger. Both of these situations will cause the computer to take too long to calculate, which is not conducive to real-time operation during particle image velocity measurement experiments.

FFT technology not only can greatly increase computation speed, but also can carry out frequency domain filtering technology on the signal, since the high frequency component is the fast change part of the image in the image processing, and the uniform background corresponds to the low frequency component. The image signal can be transformed as required by the algorithm combined with the digital signal processing tools, thus effectively improving the signal-to-noise ratio of the signal.

The correlation function matrix calculated by the above method is a two-dimensional discrete distribution. The position where the maximum value of the correlation function in the matrix is located is not necessarily the actual particle displacement value. The displacement error calculated according to the maximum correlation value is ±1/2 pixel.

In fact, higher precision displacement values can be obtained if the results obtained by these correlation functions can be further processed, which is the sub-pixel fitting algorithm.

How to obtain the pixel value corresponding to the maximum correlation value point by using the correlation value processing on integer nodes is the main direction of sub-pixel fitting research. The subpixel part of particle displacement can be determined by simulating the correlation value, and the displacement accuracy can reach 0.1 pixel. The process first finds out the integer coordinates closest to the largest phase in the region (i, j), and then extract that four points around it $(i-1, j)$, $(i+1, j)$, $(i, j-1)$ and $(i, j+1)$ Related values of R_{II} $(i-1, j)$ *Xu* R_{II} $(i+1, j)$ *Xu* R_{II} $(i, j-1)$ and R_{II} $(i, j+1)$. The displacement of particles with sub-pixels can be calculated and obtained using the fitting method of selecting three points in each direction. The fitting method Master should select an appropriate fitting function according to the contour of the correlation coefficient distribution. The Gaussian curve method has a higher fitting accuracy among the parabola method at present, the Gaussian curve method and the center of gravity method, which is mainly because the CCD camera records the particle image in the laser light, and the intensity of the particle reflected light is Gaussian distribution. It means that the gray scale distribution of the particle image taken by the CCD camera is Gauss distribution, so the distribution of the correlation coefficient is also Gauss distribution.

The Gaussian fitting function is:

$$f(x) = C \cdot e^{\frac{-(x_0-x)^2}{k}} \tag{7.23}$$

In this equation:

C Gaussian curve peak height;
X Gaussian curve peak position;
k Half width.

The coordinate fitting equation is:

$$x_0 = i + \frac{\ln R_{(i-1,j)} - \ln R_{(i+1,j)}}{2\ln R_{(i-1,j)} - 4\ln R_{(i,j)} + 2\ln R_{(i+1,j)}}$$
$$y_0 = j + \frac{\ln R_{(i,j-1)} - \ln R_{(i,j+1)}}{2\ln R_{(i,j-1)} - 4\ln R_{(i,j)} + 2\ln R_{(i,j+1)}} \quad (7.24)$$

In the equation:

i, j the pixel position with the maximum correlation value;
x_0, y_0 the pixel position after sub-pixel fitting.

In addition, many practical auxiliary algorithms such as error vector correction, pre-bias algorithm and window deformation iterative algorithm have also been developed, so I will not repeat them here.

Particle Image Velocimetry (PIV) uses three-dimensional velocity reconstruction method and adopts two cameras in Scheimpflug optical arrangement (Fig. 7.27). The stereo-particle image velocimetry (SPIV) technique is very useful for the study of unsteady and complex flow. The stereo-particle image velocimetry (SPIV) can be used to capture the illuminated area in the flow field from two off-axis directions simultaneously and record the particle image, thus obtaining a stereo-particle image velocimetry (SPIV) in a cross-section. SPIV can not only obtain the out-of-plane motion information of particles, but also eliminate the inevitable projection error in 2D-PIV, and make up for the limitations of two-dimensional measurement.

The basic method of SPIV is to use two cameras to photograph the illuminated area in the flow field from two off-axis directions at the same time according to

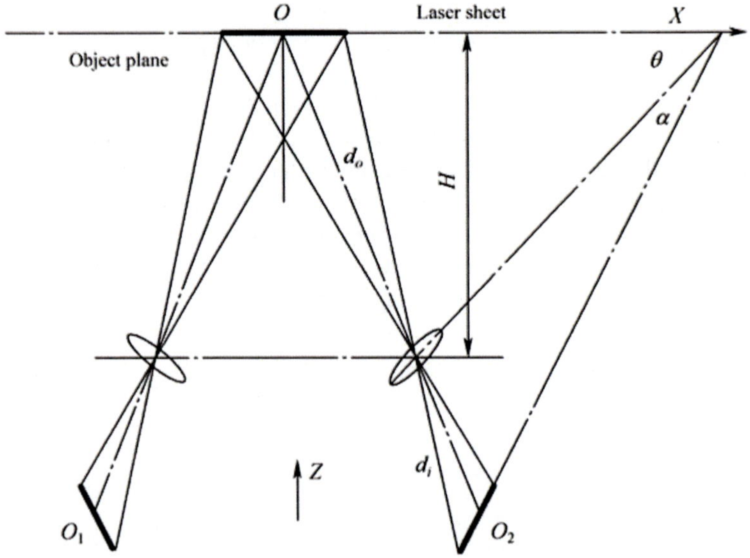

Fig. 7.27 Scheimpflug optical arrangement

Scheimpflug optical arrangement. Three-dimensional velocity reconstruction can be established by establishing the overdetermined equation without the need for geometric parameters of the system after transforming the image plane displacement determined by the two cameras to the corresponding object plane.

The three-dimensional rate reconstruction method refers to the process of transforming the image plane displacement determined by two cameras to the corresponding object plane and combining them to obtain three-dimensional information. The reconstruction method is divided into geometric reconstruction method and calibration reconstruction method. The calibration reconstruction method is divided into two-dimensional calibration reconstruction method and three-dimensional calibration reconstruction method. Nowadays, the calibration reconstruction method is often used for calibration.

The three-dimensional calibration reconstruction method gets rid of the dependence on system parameters compared with the two-dimensional calibration reconstruction method. The main realization method is that the calibration plate is not only shot at a fixed Z, but also moves the calibration plate several times in the optical domain to obtain the object spatial position data, and obtains the corresponding relationship of the object spatial coordinates on the image plane through the corresponding algorithm, thus reconstructing the displacement information in three directions. The calibration board should be placed in the test area when actually calibrating the calibration board, must be consistent with the sheet light, and should include the entire test range. First use two cameras to capture the image of the calibration plate after the SPIV camera layout is completed, so that the transformation function F_0 is usually used to determine the template with the same contour of the grid point on the calibration plate according to the relationship between the image plane coordinates of the calibration plate and the correspondence between the space position and the grid points in the plane are matched to obtain the position with the grid points in the image plane.

The corresponding relation between the coordinate x of the three-dimensional object space and the coordinate x of the image plane of each camera is: $X = F(x)$, for the given point of the object space (x, y, z), by Eq. (7.24) its coordinates on the image plane can be determined (X_1, Y_1) and (X_2, Y_2), i.e. that image plane displacement(ΔX_1, ΔY_1) (ΔX_2, ΔY_2) can be determine. Since $\Delta X = F(x + \Delta x) -$ F(x). It can be approximated as $\Delta X \approx \Delta F(x)\Delta x$, where $(\nabla F)_{ij} = \partial F_i/\partial x_j = F_{i.j}$, and an overdetermined equation system can be established:

$$\begin{bmatrix} \Delta X_1 \\ \Delta Y_1 \\ \Delta X_2 \\ \Delta Y_2 \end{bmatrix} = \begin{bmatrix} \frac{\partial X_1}{\partial x} & \frac{\partial X_1}{\partial y} & \frac{\partial X_1}{\partial z} \\ \frac{\partial Y_1}{\partial x} & \frac{\partial Y_1}{\partial y} & \frac{\partial Y_1}{\partial z} \\ \frac{\partial X_2}{\partial x} & \frac{\partial X_2}{\partial y} & \frac{\partial X_2}{\partial z} \\ \frac{\partial Y_2}{\partial x} & \frac{\partial Y_2}{\partial y} & \frac{\partial Y_2}{\partial z} \end{bmatrix} \cdot \begin{bmatrix} \Delta x \\ \Delta y \\ \Delta z \end{bmatrix} \tag{7.25}$$

In this equation:

x, y, z the three-dimensional coordinates of the object space position (mm);

X, Y the pixel coordinate of the photo taken by the camera corresponding to the object space.

The three-dimensional displacement (Δx, Δy, Δz) can be determined by Eq. (7.25).

7.3.2 Particle Image Velocimetry System

A typical PIV system includes a tracking particle spreading system, an illumination system, an image Acquisition system, and a synchronous Control system.

First introduce the tracking particles that capture the flow field information and the corresponding lighting system. The existence of tracer particles in the studied fluid is essential for PIV technology, which can reflect the real flow field information. The basic requirement of PIV technology for tracer particles is that [29] It should have good following performance, scattering sufficient light intensity under experimental conditions to obtain satisfactory signal-to-noise ratio of the detector, appropriate concentration in the fluid to obtain continuous time and space information, and the particles should be non-toxic, non-corrosive, non-abrasive and stable in chemical properties. According to Stokes theory [30]. The related parameters of particle followability obtained:

$$\tau = \frac{d_p^2}{18\upsilon_f}\left(\frac{\rho_p}{\rho_f}\right) \tag{7.26}$$

In this equation:

τ time constant;
ρ_p particle density (kg/m^3);
ρ_f fluid density (kg/m^3);
d_p particle diameter (μm).

τ represents the speed of the particle reaction. It can be seen from this conclusion that the time constant τ is related to ρ_p (particle density), ρ_f (fluid density), and d_p (particle diameter). The smaller the ρ_p and the closer to the density of the fluid medium, the smaller τ.

Generally, the density of tracer particles is relatively high. In this case, the size of *D* can be Coerced as long as the particle size is small enough. It is found in practical applications that it is more difficult to perform high-quality particles injection in the gas flow field compared with the liquid flow field application. It is necessary to ensure a better tracking of the flow field movement since the particle density is quite different from the flow field density. The diameter of the tracer particle must be small enough.

The light scattering properties of particles are closely related to the ability to obtain high-quality images. It is more economical and effective to select particles with good scattering properties than to increase the laser intensity to improve image quality. In general, Mie scattering theory can be applied to discuss and analyze them for spherical particles whose particle size is larger than the wavelength of incident light [31]. Mie scattering can be characterized by a standardized diameter, q is defined as:

$$q = \frac{\pi d_p}{\lambda} \tag{7.27}$$

In this equation:

q standardized diameter;
λ wavelength of incident wave (nm).

Figures 7.28 and 7.29 show the polar intensity distribution of scattered light of particles of different sizes in the air and water respectively under the irradiation of 532 nm wavelength light according to the Mie scattering theory, in which the intensity is calibrated on logarithmic scale, and the difference between adjacent scales is 100 times.

It can be seen that as the particle diameter increases, the scattered light intensity has an obvious trend of enhancement, and as the observation angle increases (0–180°), the scattered light intensity also increases. However, the most commonly used observation angle is 90° in the actual analysis of Slaved. As a result, the scattered light intensity as a function of particle size presents a rapidly oscillating identity. On the other hand, the scattering efficiency strongly depends on the ratio of the refractive index of the particles to the refractive index of the fluid. Larger particles are needed for experiments in water since the refractive index of water is higher than that of air.

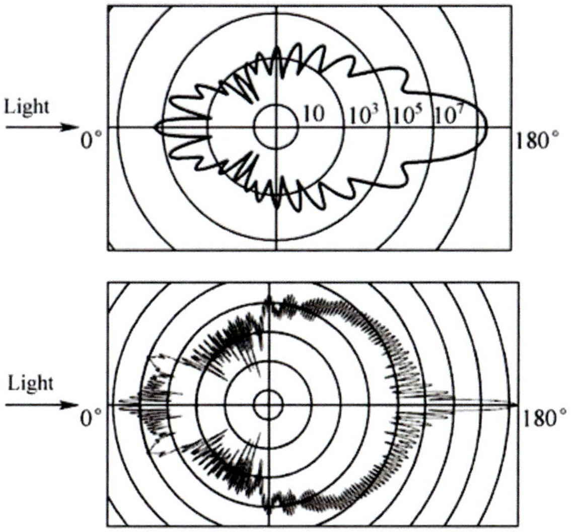

Fig. 7.28 Micron (left) and 10 μm (right) scattered light of oil particles in the air

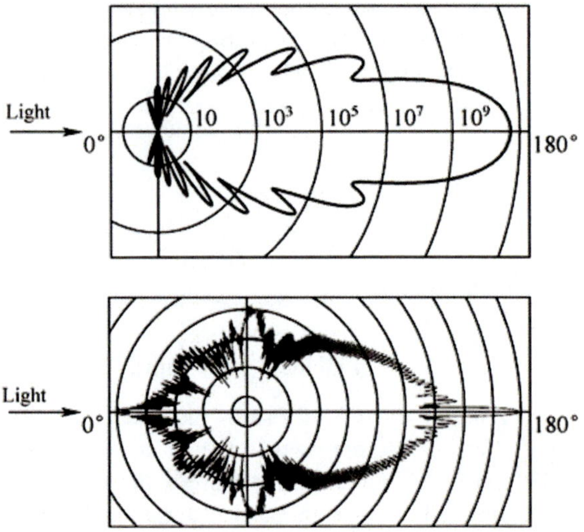

Fig. 7.29 Micron (left) and 10 μm (right) scattered light of glass particles in water

Table 7.5 Particle materials commonly used in liquid flow field [32]

Types	Material	Average size/μm
Solid	Polystyrene	10–100
	Aluminum	2–7
	Glass ball	10–100
	Synthetic coating pellets	10–500
Liquid	Several kinds of oils	50–500
Gas	Oxygen bubbles	50–1000

See Tables 7.5 and 7.6 for commonly used particle materials in liquid flow field and gas flow field.

Appropriate particle generation and spreading technology can ensure the proper concentration and continuity of particles in the fluid. There are many technologies used to produce and supply particle in the gas flow field [33]. For example, dry

Table 7.6 Particle materials commonly used in gas flow field [32]

Types	Material	Average size/μm
Solid	Polyethylene	0.5–10
	Aluminum	2–7
	Magnesium	2–5
	Glass microsphere	30–100
	Dicarboxylic acid	1–10
Liquid	Several kinds of oils	~1
Gas	–	0.5–10

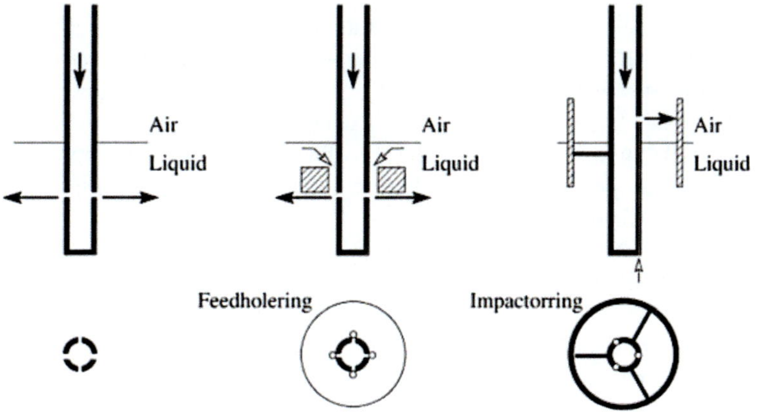

Fig. 7.30 Basic forms of Laskin nozzle

powder can be sprayed with a fluidized bed gas nozzle. The liquid can be evaporated and then cooled by a device called a cooling generation. Obtained through a special atomizer, the atomizer can also be used to spread solid particles floating in the evaporated liquid, or for oil such as generation fine droplets. Cooling generators, smokers, polystyrene dispensers or latex particles injected with water and alcohol are often used for flow field visualization and LDV in the early stage for the wind tunnel flow field. Laskin nozzle, particle generator and oil are used for PIV measurement in gas flow field. These particles have the advantages of non-toxicity, good scattering, no obvious deformation in air for several hours and in various states. They can also be used as tracer particles in the flow field of the whole wind tunnel or local injection particles for a certain flow tube in circulating wind tunnels. The injection method of particles mostly uses perforated plates or injection rake injection, which are specifically designed according to the corresponding experimental environment [34].

Laskin nozzle (Fig. 7.30) is a relatively simple nozzle with relatively simple structure, which can generate high concentration poly dispersed submicron aerosol particles and has been adopted by many foreign aerodynamic research units as DPIV oil mist tracer particle generator. Laskin nozzle is a porous nozzle, and its principle of aerosol generation is very simple, that is, compressed air is ejected from small tubes or orifices at a very high speed. Liquid is sucked up, entered into high-speed airflow, fragmented and dispersed to form aerosol with smaller particle size due to Bernoulli effect, negative pressure is generated in the outlet area. The raw materials for generating aerosol can be di-iso-octyl sebacate (DEHS), olive oil, ethylene glycol, etc. Glycol particles have better light reflection, but the particles are not as small as olive oil and DEHS, and the particle uniformity is also poor. Olive oil particles are small, but their reflectivity is relatively weak. DEHS particles are small, uniform in size and strong in luminescence. In modern times, DEHS is generally used as the raw material for preparing tracer particles.

China Academy of Aerospace Aerodynamics has designed three different Laskin nozzles (I, II, III) after investigation and research, and tested the three nozzles under different inlet pressures to analyze the particle size distribution and concentration changes of aerosol generated by them. The test results are shown in Table 7.7.

The Laskin nozzle was used as the core device to start the development of a large oil mist particle generator after the Laskin nozzle was developed (Fig. 7.31). The oil mist particle generator consists of a gas distribution system, a pressure regulation and control system, an oil mist generation main box, a residual liquid collection chamber, an oil supply and discharge and oil level indicating device, an oil mist particle outlet group, a control panel and other parts. As for that nozzle and oil in the oil mist generate main box body, the oil mist generator has 96 group of nozzles to ensure the generation of sufficient flow oil mist particles since 12-way primary air separation system and 8-way secondary air separation system are adopted, and each inlet of the primary air separation system is provided with pressure reducing valves to realize the inlet pressure from 0 to 0.4 MPa is accurately adjustable. It is only necessary to close the inlet valve behind the pressure reducing valve of each branch, open the air flow main valve, adjust the inlet static pressure to the required value, and open the inlet valve of each branch when particles are required in the experiment. The pressure reducing valve for the same particle requirement can be adjusted and the air flow main valve can be operated to open and close.

Table 7.7 Laskin nozzle test results

Nozzle I			
Absolute pressure/MPa	0.15	0.20	0.25
Median particle size/nm	130.7	122.3	120.2
Nozzle H			
Absolute pressure/MPa	0.15	0.20	0.25
Median particle size/nm	122.1	116.4	112.4
Nozzle M			
Absolute pressure/MPa	0.15	0.20	0.25
Median particle size/nm	183.4	165.7	135.8

Fig. 7.31 Large oil mist particle generator and particle spreading pipe rack

Oil mist tracer particles have better followability under low-speed and subtransonic experimental conditions due to their low density and small diameter. However, for hypersonic wind tunnel experiments, its particle size on the order of micron is relatively large, and it is often attached to the wall of the wind tunnel at the throat of the nozzle, causing pollution to the wind tunnel, and its follow-up performance is poor when strong shear flow and shock wave pass through, resulting in its limited use in hypersonic wind tunnel experiments. The nano-solid powder tracer particles are mostly oxide powder particles of generation through chemical synthesis and liquid phase synthesis. Their primary particle size is in the range of tens of nanometers to tens of nanometers. They have excellent followability and are suitable for high Supersonic wind tunnel experiment.

However, it is easy to agglomerate and is difficult to disperse due to the larger surface energy and strong polarity of the nano-solid powder. The diameter of the agglomerated solid powder is on the order of micrometers and may be larger. The inertia of the solid powder micelles is larger and the followability is poor since the diameter of the micelles is too high, causing a series of problems. For example, solid powder micelles cannot enter the boundary layer, there is slippage in the strong shear air flow, and there is a hysteresis phenomenon when passing through the shock wave structure to smooth out the shock wave structure, and the PIV measurement results will produce inevitable errors. Therefore, how to use a certain method to crush the agglomerated solid powder clusters and injection into the experimental section is the key to the application of nano-solid powder tracer particles to the hypersonic wind tunnel PIV experiment.

To solve the problem of agglomeration of nano-solid powder tracer particles, the method of high-speed versus collision and cyclone separation can be used to break up particle clusters through collision. High-speed collision is to use the huge kinetic energy bred by the high-speed airflow generated by the high-pressure gas through the nozzle to make the particle microclusters collide with each other or with the fixed plate (such as the impact plate) to achieve the purpose of crushing. The cyclone separation process causes the solid powder to rotate in the separation chamber driven by the high-speed gas and form a strong rotating airflow, which accelerates and mixes the particles and causes impacts and collisions. The fine solid powder qualified for pulverization is pushed to the center of the cyclone separation chamber by the air flow, while the coarser solid powder continues to circulate in the separation chamber and continuously participates in the pulverization process, thus achieving the purpose of pulverization and realizing screening.

Based on research and test experience, China Academy of Aerospace Aerodynamics has organically combined fluidized bed, high-speed collision, cyclone separation and other functions to design a new nano solid powder generator, as shown in Fig. 7.32.

The nano solid powder generator (Fig. 7.32) uses a high-pressure gas source, rectifies it through a porous plate and a honeycomb device, enters a fluidized bed, realizes a boiling state, drives particles to pass through a layer of high-mesh screen, enters a nozzle to accelerate, carries out high-speed collision, and realizes airflow pulverization through collision and shearing. Then enter the separation chamber for

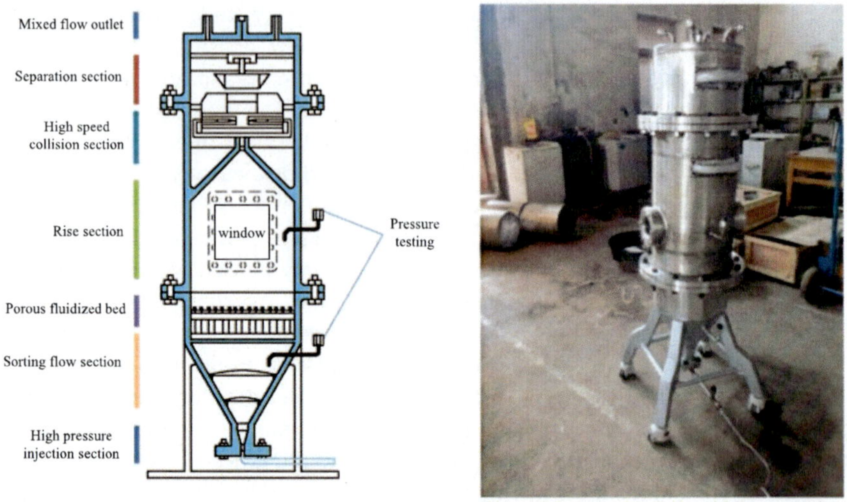

Fig. 7.32 Schematic diagram and physical diagram of nanometer solid powder generator

cyclone separation. Fine particle powder is output by airflow through the upper outlet and enters the experimental test area. However, the coarser solid powder continues to circulate in the separation chamber and continuously participates in the crushing process, thus achieving the purpose of crushing and realizing screening.

PIV illumination system has been relatively mature in the development stage of laser speckle velocimetry technology. The light source, light path and beam expansion method adopted will change according to different specific research objects, but the basic principles are the same. The commonly used light sources are pulsed lasers, such as Nd: YAG (neodymium yttrium aluminum garnet) lasers or ruby lasers. The 532 nm laser beam is expanded with a cylindrical lens to form a sheet light source to illuminate the flow field. Some people also use a continuous laser light source, such as an argon ion laser. The pulse interval for acquiring images is provided by a chopper or modulator Coerced. The internal structure of a typical double exposure laser is shown in Fig. 7.33.

Early particle image recording was done by exposing the particle image to the film. The silver halide particles fixed by gelatin acted as a photosensitive emulsion, interacting with incident light to record the particle image information. The development process needs to be carried out in a darkroom, and related characteristic parameters such as spectral sensitivity, rate, resolution, gray scale and noise, etc. need to be carefully considered. Although it seems that this process is quite cumbersome, in the early days, Adrian, Keane, and Yao [35–37] conducted a lot of PIV statistical evaluation work on the particle images recorded on versus film, and gave a detailed mathematical description that is developed from autocorrelation analysis to exchange analysis, and applied to experimental research [38, 39].

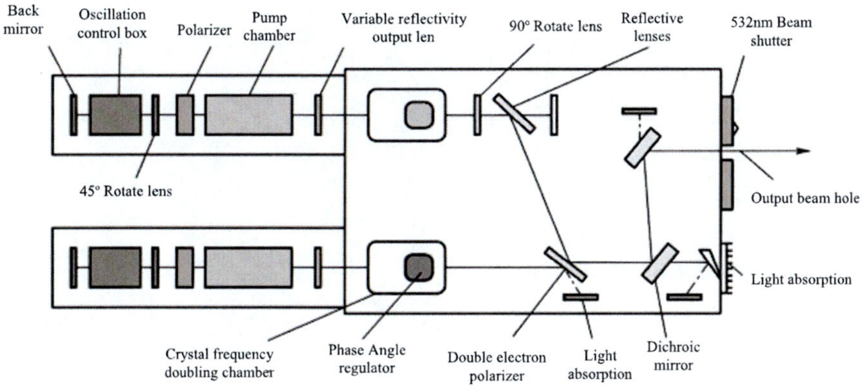

Fig. 7.33 Internal structure diagram of typical double exposure laser

The digitization of particle image records began with Willert and Gharib [40] abandoning the traditional photographic film and using CCD video camera to record particle video images, this recording method gets rid of the time-consuming and laborious methods of film development and electro-optic mechanism analysis, and promotes the process of widely adopting Digital PIV.

Westemeel [41, 42] proposed a DPIV system with the same resolution as the PIV system using film recording, and also analyzed and summarized the corresponding estimation methods of the DPIV system. Then the researchers [43, 44] continued to improve the resolution, accuracy, accuracy of the DPIV system and analytical method.

The emergence of cross-frame cameras makes it possible to capture images with extremely short exposure intervals. It is different from ordinary cameras in that it uses a special CCD chip. Each pixel on the photosensitive array of a cross-frame CCD camera is composed of an adjacent photosensitive area and a shielded non-sensitive area (Fig. 7.34). As a buffer, the signal generated by the exposure of the photosensitive area is instantly stored after the camera shutter is opened, so that the photosensitive area can quickly be exposed for the next frame of image. As a result, the cross-frame camera not only realizes that the two light pulse exposures are recorded on two consecutive frames of images, but also ensures that the time delay (gold) of the two light pulses can be adjusted. The current time delay can be as small as 200 ns, which is satisfactory for the vast majority of hypersonic flow field measurement needs.

A synchronization Control system is required to synchronize the various mechanical and electrical signal, in order to synchronize the pulse laser with the Acquisition image of the CCD camera, or to synchronize the change of the experimental phenomenon with the acquisition image (Fig. 7.35). In the test, there is generally a reference period signal as the synchronous signal of various instruments, and the signal of other instruments start to work with the delay signal of this reference signal as the trigger. The delay signal generator for typical Nd: YAG laser and cross-frame

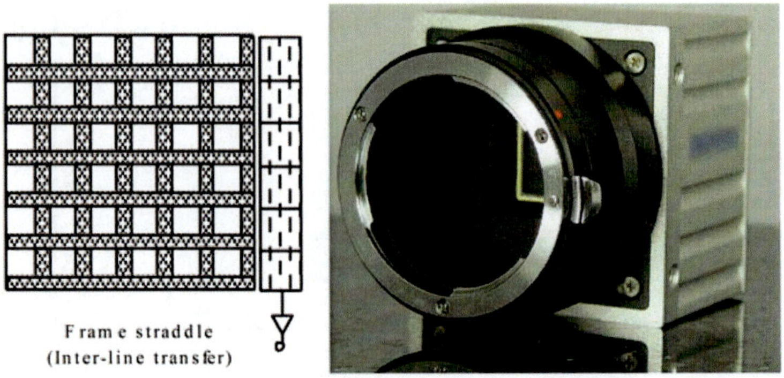

Fig. 7.34 Schematic diagram of cross-frame CCD array types

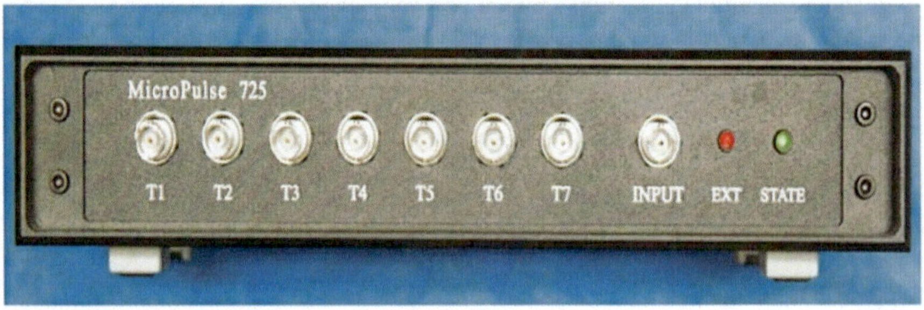

Fig. 7.35 MicroPulse725 delayed synchronous coerced

single cross-frame CCD camera needs to output at least 5 delay signals to control the synchronous operation of the laser and CCD. For example, the typical Micro Pulse725 delay signal generator can output 8 channels of delay Signal, using standard TTL signal format, can be externally triggered to lock the work, delay accuracy 0.25 ns, using computer software Coerced.

Specifically, the Coerced generator is synchronized to generate the cycle signal during the experiment, and the xenon lamp (F_1, F_2) of the 2 lasers and the Q-switched Coerced signal (Q_1, Q_2) are output through the internal delay. At the same time, the CCD camera trigger signal T_{ri} is output (Fig. 7.36). In addition, in order to activate the synchronous trigger T_{ri} of the Q switch Q_1 and the CCD digital camera at the same time during double exposure, there is also a Q_1, Q_2, and T_{ri} synchronous switch Coerced.

In the past 20 years, PIV technology has gone out of the experimental research stage, developed from the original two-dimensional PIV using film to record particle images to the present two-dimensional digital PIV and three-dimensional digital PIV systems, successfully applied to fluid mechanics and related research fields, and began to serve the engineering practice [45]. At present, some commercial PIV

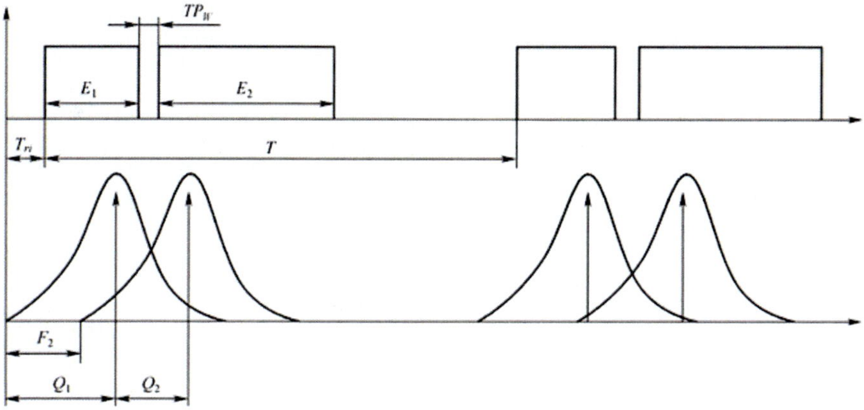

Fig. 7.36 Q Nd: YAG Laser delay control schematic diagram

companies such as TSI in the United States, Dantec in Denmark and LAVISON in Germany have all launched two-dimensional and even three-dimensional PIV complete sets of equipment for the majority of experimental researchers to choose.

7.3.3 Engineering Application Examples of Particle Image Velocimetry Technology

7.3.3.1 Application of DPIV Technology in Flow Field Measurement Experiment of Flying Wing Model in Large Low Speed Wind Tunnel

(1) Experimental conditions and layout

The flow field testing technology in the production wind tunnel is established according to the layout requirements of the new generation of aircrafts to provide technical support for the development of advanced layout aircrafts. For the standard model of the new layout aircraft, DPIV space flow field measurement test technology is used in FD-09 wind tunnel to obtain the flow field test results of typical cross sections (Fig. 7.37). Table 7.8 shows the combined working conditions such as the station position, incoming flow velocity and angle of attack measured by the test. PIV flow field measurement is carried out for these working conditions.

The FD-09 wind tunnel is a large low-speed recirculation wind tunnel of the order of 3 m × 3 m. Its incoming flow rate is continuously adjustable from 120 to 100 m/s. There are observation windows on both sides of the wind tunnel, with its own angle of attack and side slip angle mechanism. Low-speed wind tunnel experiments with larger models can be carried out (Figs. 7.38 and 7.39).

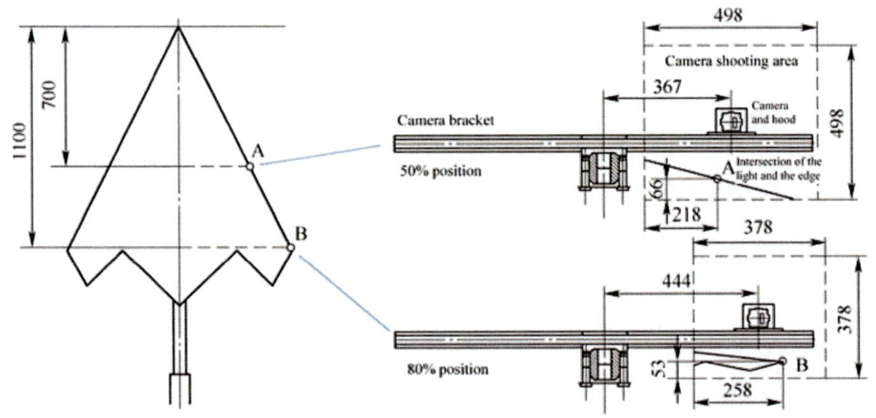

Fig. 7.37 Camera shooting position (The picture above shows the 50% station position, and the picture below shows the 80% station position. When the actual angle of attack changes, the Y-direction position of reference points A and B may change, and the X-direction position remains unchanged)

Table 7.8 Experimental conditions of flying wing standard model in FD-09 wind tunnel

Position/%	Incoming flow rate/(m/s)	Angle of attack/(°)
50	50	4, 6, 12, 16, 20, 28, 42
80	50	4, 6, 12, 16, 20, 28, 42

Fig. 7.38 Model, camera bracket and spread tube

(2) Experimental results and analysis

As can be seen from that velocity vector (Fig. 7.40) and the mean vorticity diagram (Fig. 7.41) of the PIV experiment, leading edge vortices begin to appear on the wing when $a = 6°$, but the vorticity is small and no obvious vortex core appears. The vorticity at the vortex center increases obviously when $a = 12°$, which indicates that a concentrated vortex core has been formed at this time. Compared with the force measurement results, it can be seen that the formation of concentrated vortex

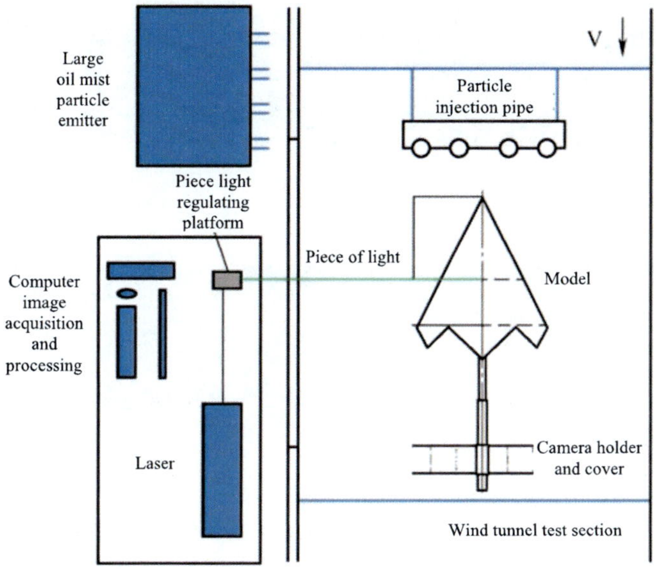

Fig. 7.39 Experimental layout diagram

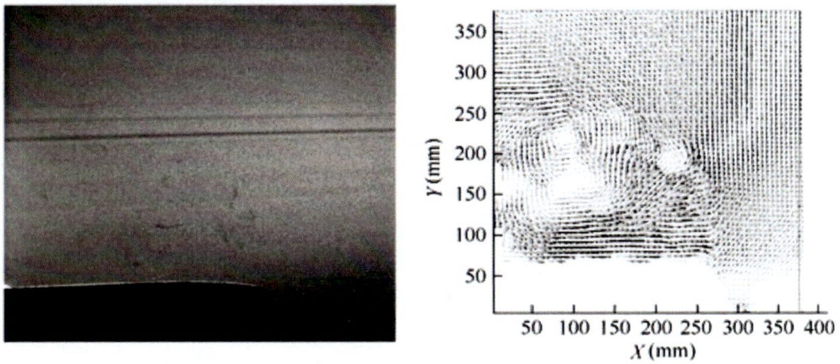

Fig. 7.40 80% station, angle of attack = 20 Time particle image and instantaneous rate vector distribution image

produces nonlinear vortex lift. The vortex core at 50% of the full-length section gradually expands and tends to move to the symmetry plane of the body when the angle of attack increases from 12° to 20°. The vorticity at the vortex center decreases obviously although the vortex area at 80% of the full-length cross section gradually expands, which indicates that the vortex core has broken from the nose to the tail. It has a great influence on the pitching torque combined with the analysis of force measurement results, since the leading-edge vortex breaks at the tail.

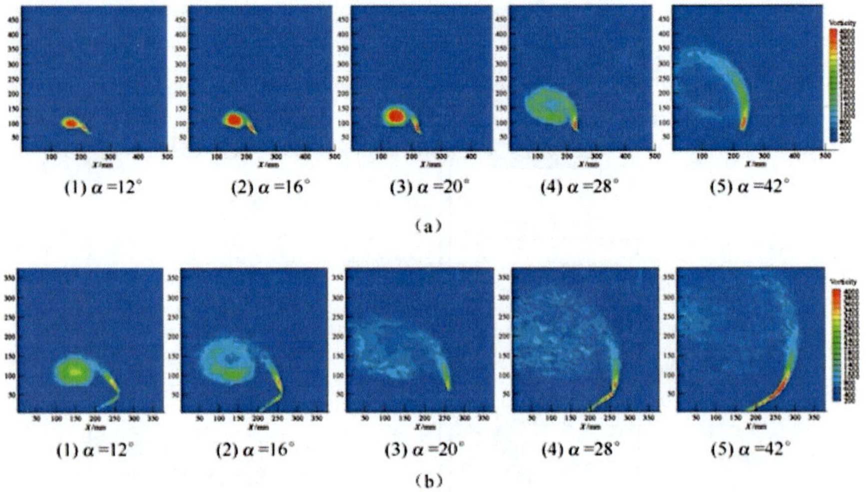

Fig. 7.41 Average vorticity chart for PIV experiment. **a** 50% plane vorticity distribution diagram of full length; **b** 80% full-length plane vorticity profile

The rupture point of the vortex core gradually develops toward the nose as the angle of attack further increases, when the angle of attack is 28°. At this time, the vortex core at 50% of the full-length section has broken, and the vortex at the vortex core is significantly reduced, but there is still circulation around the vortex core; there is no obvious circulation around the vortex core at 80% of the full-length section at this angle of attack. The present state is a small-scale vortex after the vortex core ruptures. There is no obvious vorticity on the front and rear cross-sections when the angle of attack is 42°, indicating that the vortex has been completely broken at this time, combined with the force measurement results, a stall has occurred at this time.

Figure 7.42 is the instantaneous rate vector diagram of the PIV test. Three instantaneous rate vector diagrams are given at each angle of attack. Observing 80% of the full-length cross section, it can be seen that the vortex core is basically stable with time when $a = 12$, and the whole vortex presents a conical shape from the front to the back. The vortex core begins to swing left and right when $a = 16°$. When $a = 20°$, the vortex core has already started to rupture and several vortices have appeared. The distribution of vorticity is unsteady.

In general, the rupture process of the vortex is gradually from the stable contour of the vortex core to the vortex core begins to swing, and finally the vortex breaks.

7.3.3.2 Application of DPIV Technology in C919 High Speed Head Selection Test

Parabola or symmetrical airfoil head should be rotated to form the nose shape, and streamline should be properly carried out in the aerodynamic layout design of modern

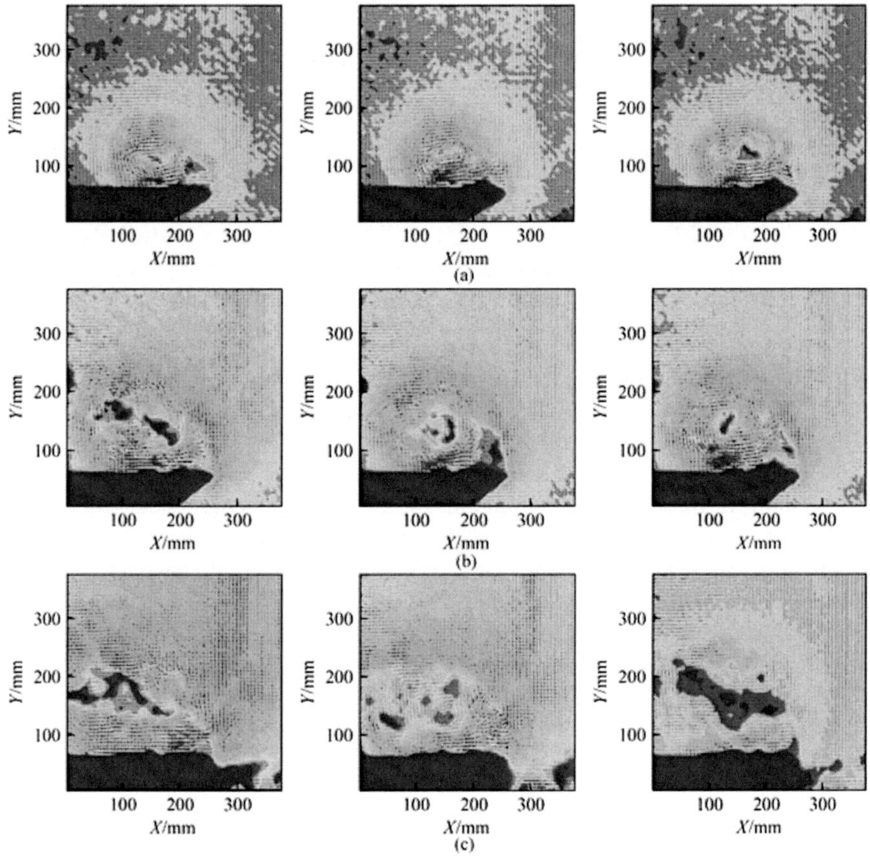

Fig. 7.42 Vector diagram of instantaneous velocity in PIV experiment. **a** $a = 12$ 80% full captain position; **b** $a = 16$ 80% full captain position; **c** $a = 20$ 80% full captain position

civil aircraft nose. The resistance should be small to avoid separation and increase the resistance in the aerodynamics. Less straight lines to reduce the supersonic wave resistance and avoid shock waves in the porthole position of the nose during high-speed cruising, which will affect the comfort of the cockpit. The high-speed wind tunnel test task for the nose of the C919 aircraft was proposed by Commercial Aircraft Corporation of China in July 2010. The purpose of the test is to simultaneously use the three test methods of nose pressure measurement, oil flow display and DPIV measurement. Flow field characteristics of different design schemes. The advantages and disadvantages of different design schemes are compared through the flow field data of different heads obtained by the above three test techniques, so as to select the best design head scheme.

For a specific 1.2 m sub-transonic wind tunnel (FD-12), the goal of this project is to solve the problems of preparing and spreading large-flow tracer particles, optical path arrangement and equipment vibration reduction, and finally realize the coordinated operation of DPIV system and wind tunnel operation.

The application of DPIV technology Master in FD-12 has the following difficulties:

(1) The flow rate of large-scale sub-transonic wind tunnels is much higher than that of ordinary low-speed wind tunnels and small hypersonic wind tunnels. The tracer particles fill all the experimental space when the wind tunnel is running.
(2) There is an optical window on the side, but the top optical window is inside the chamber. There is a porous plate between the chamber and the experimental section. There is unstable airflow and vibration interference in the chamber.

The following technical schemes are adopted to establish particle image velocimetry technology suitable for 1.2 m sub/transonic wind tunnels in order to solve these problems:

The statistical study of particle diameter and quantity concentration is carried out according to the existing small atomization particle generator with slot nozzle to obtain the data of particle following characteristics, concentration change with inlet pressure, etc. A large, atomized particle generator group, auxiliary air sources and pipelines are designed according to the demand of large flow of atomized particles in sub/transonic wind tunnel. Aiming at the unstable airflow and vibration interference in the narrow resident space, a vibration damping platform suitable for its size and a mobile platform group suitable for sheet light source are designed, which provides the possibility for the realization of experimental light path.

(1) Experimental conditions and layout

The model used in this test is provided by the design department. The model scale is 1:30. It is a light fuselage structure (without wings, tails and wing body rectifiers). The configuration definition is: light fuselage two nose +The straight section of the fuselage + the tail section of the fuselage. The model is not limited by the wind tunnel observation window in the nose pressure measurement and oil flow display tests, and an all-optical fuselage structure model is adopted. The basic dimensions of the model: total length, 1296.67 mm; the cross-sectional area of the straight section of the fuselage, 0.01435 m^2; the maximum width of the fuselage, 131.99 mm; the maximum fuselage Height, 138.87 mm.

The model is cut off at a distance of 670 mm from the nose in the DPIV measurement experiment in order to ensure that the nose is in the wind tunnel observation window of the position and to ensure that the model and the wind tunnel tail support mechanism have sufficient distance. This scheme can ensure that the slenderness ratio of the model is about 5, so as to minimize the influence of model truncation on the original flow field of the nose. The design provides three heads, Model-1, Model-4 and Model-5 respectively. Model-1 and Model-5 were tested in DPIV test. On the model, there are 182 pressure measuring holes in the direction of the downstream line, that is, the direction of the head of the model (Fig. 7.43). 12 pressure measuring sections are arranged along the streamline direction on the bus bar of the

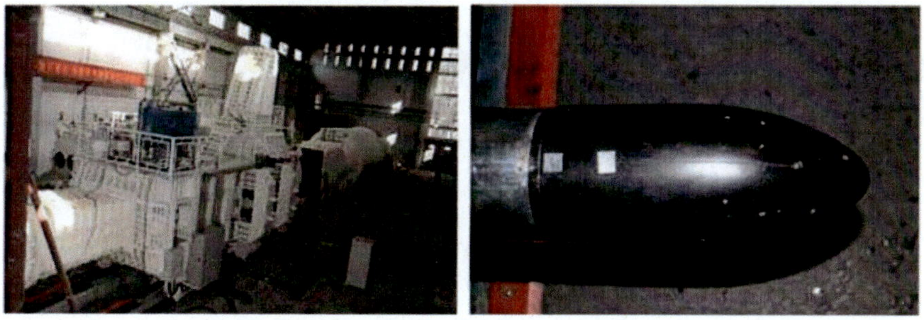

Fig. 7.43 1.2 m, supersonic wind tunnel and nose test model

leeward symmetry plane of the nose, and a pressure measuring point is arranged at the leading edge point of the nose, i.e. the zero section, to measure the stagnation point pressure. The test Mach number is 0.785, 0.82, the angle of attack is 3°, 6°, a total of 8 working conditions.

The optical damping platform was set up in the chamber above the perforated plate of the experimental section before the experiment. The airbag pressure was 0.3 MPa. The laser was fixed on the platform. The laser exit hole was equipped with a 0.6 m long light guide arm, and a multi-dimensional light guide arm was fixed on the platform. The moving system (three-dimensional translation, one-dimensional rotation, one-dimensional deflection) is at the front end of the laser. The integrated sheet light source on the light guide arm is fixed on the multi-dimensional mobile system. The laser sheet light passes through the optical glass window on the perforated plate to illuminate the flow field. The CCD camera takes pictures of the flow field through the side window (Fig. 7.44).

A calibration board was first placed in the experimental area and photographed with a CCD camera before the experiment to determine the actual measurement area size and the scale relationship between the CCD camera pixels and the calibration board size. The approximate flow velocity of the experimental area can be determined from the total temperature of the incoming flow and the Mach number of the experimental area. The laser pulse exposure interval can be determined according to the actual size of the space represented by the CCD camera pixels and the planned displacement pixel number.

First turn on the particle broadcaster (Fig. 7.45) during the experiment, and then run the wind tunnel to shoot with the CCD camera.

Fluorescent dyes are also applied to the model, and a narrow-band color filter ((532 ± 5) nm) is installed on the CCD camera in order to eliminate the influence of the reflected light from the model. The optical lens of the CCD camera is Nikkon 50 mm/F1.4, the exposure interval of the CCD camera is 4 μs, the size of the shooting area is 400 mm, and the thickness of the laser sheet is 1 mm.

The rate field near the nose can be measured through the DPIV experiment (Fig. 7.46). The Mach number distribution cloud map near the nose can be calculated based on the adiabatic assumption (the boundary layer can be considered as an isentropic flow) if the total temperature of the incoming flow is known.

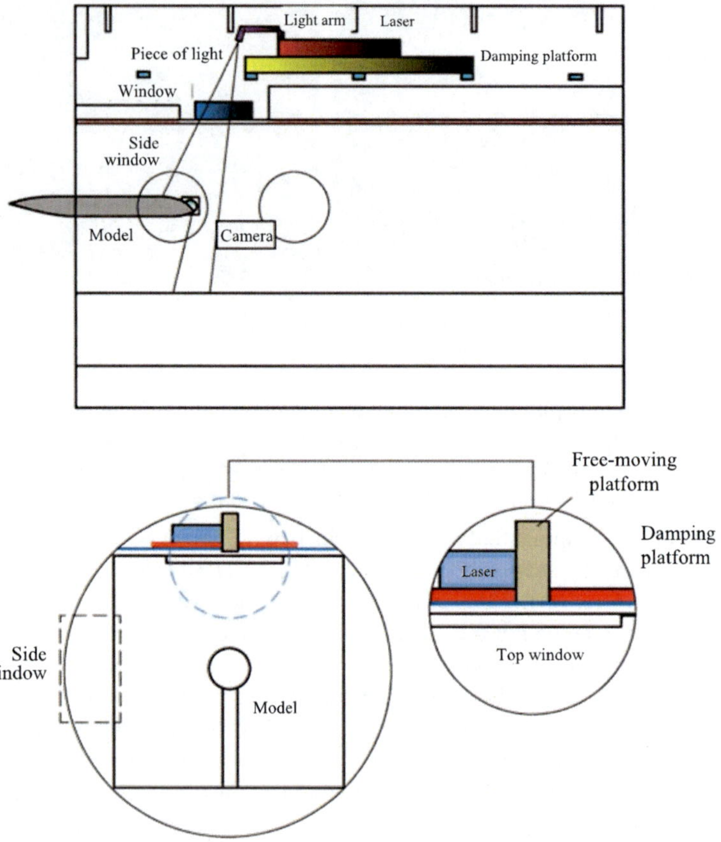

Fig. 7.44 Experimental optical path layout

Fig. 7.45 Large flow atomize particle generator system, erection of indoor vibration damping platform and external CCD camera

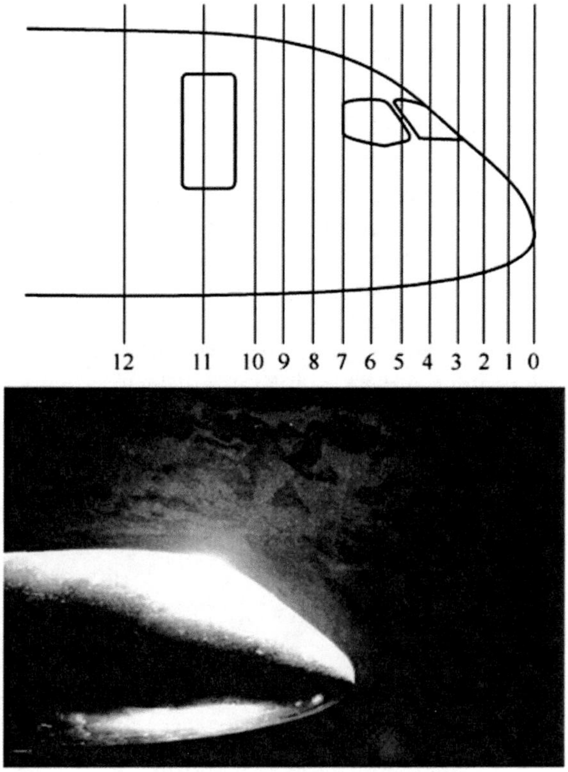

Fig. 7.46 Tracer particle image samples

$$Ma = \frac{V}{C} = \frac{V}{\sqrt{\gamma R\left(T_0 - \frac{(\gamma-1)V^2}{2\gamma R}\right)}} \tag{7.28}$$

In this equation:

Ma Mach number;
γ specific heat ratio;
R General gas constant (J/kg);
T_0 Total temperature of incoming flow (K).

This test provides the vector field distribution of velocity field near the nose. The velocity field vector field obtains two interpretation areas of the same size at the same position on the acquired front and rear two frames of particle images. Cross-correlation algorithm and fast Fourier algorithm are used to obtain the average displacement of tracer particle swarm in the interpretation area, the average displacement is corrected by sub-pixel fitting method, and then the absolute velocity vector value can be obtained according to experimental parameters. A multi-scale iterative algorithm and a deformed window algorithm are also used in the actual image transactions, in order to improve the calculation accuracy and rate field resolution.

(2) Experimental results and analysis

Figures 7.47, 7.48, 7.49 and 7.50 is the DPIV test result of Model-1. Judging from the velocity field and Mach number distribution, it gradually starts to decelerate when the high-speed airflow approaches the nose and decelerates to zero near the leading edge of the nose to form a stagnation point. This point should be the area with the maximum pressure, and then the airflow follows the symmetry plane of the nose (90° Bus, leeward symmetry) flows upward and gradually accelerates. The airflow enters the high-velocity area ($0.95 < Ma < 1.0$, abbreviated below) on the 8# pressure measuring surface, and the flow velocity is close to the speed of sound when a =

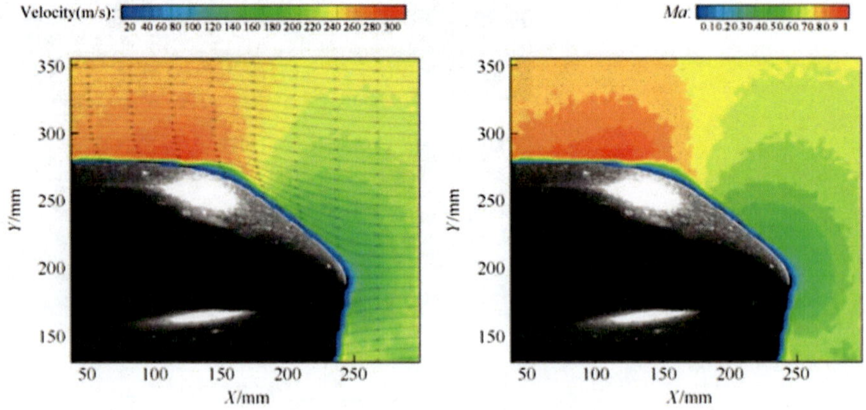

Fig. 7.47 Model-1, $a = 3°$, $Ma = 0.785$ streamline and mach number distribution

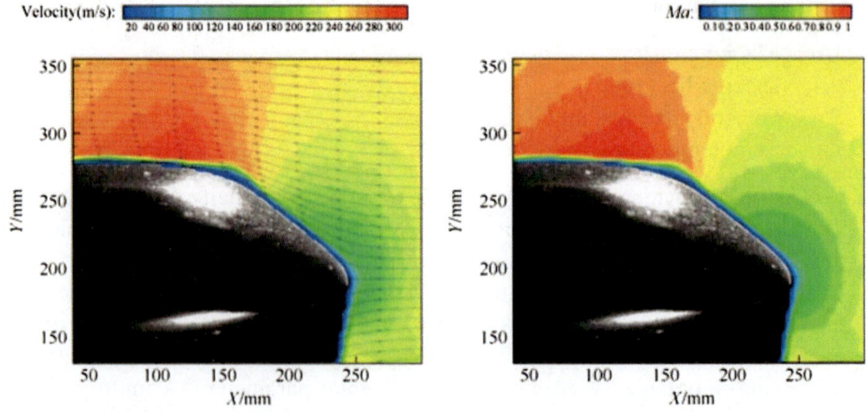

Fig. 7.48 Model-1, $a = 3°$, $Ma = 0.820$ streamline and mach number profile

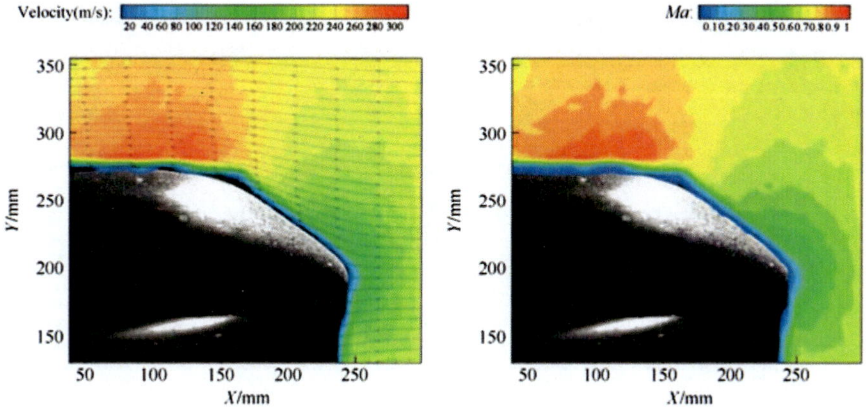

Fig. 7.49 Model-1, $a = 6°$, $Ma = 0.785$ streamline and mach number distribution

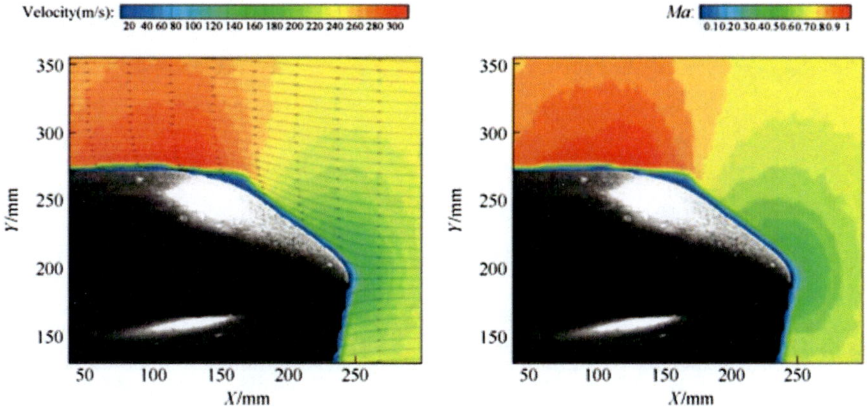

Fig. 7.50 Model-1, $a = 6°$, $Ma = 0.820$ streamline and mach number distribution

3°, $Ma = 0.785$, but does not exceed the speed of sound. The airflow enters the high-velocity area on the 7# pressure measuring surface, and enters the supersonic velocity area on the 8# pressure measuring surface ($Ma \geq 1.0$, slightly below) when $a = 3°$, $Ma = 0.820$, and the supersonic velocity area is very small. The airflow crosses the 7# pressure measuring surface and enters the high-velocity area when $a = 6°$, $Ma = 0.785$. The velocity is close to the speed of sound, but does not exceed the speed of sound. The airflow enters the high-velocity zone between the 6# and 7# pressure measurement surfaces when $a = 6°$, $Ma = 0.820$, and the airflow crosses the 7# pressure measurement surface and enters the supersonic zone, which is very small.

Figures 7.51, 7.52, 7.53, 7.54, 7.55 and 7.56 shows the DPIV test results of model-5. The experimental results of model-5 are very similar to those of model-1, but there are some differences. The airflow enters the high-velocity area between the

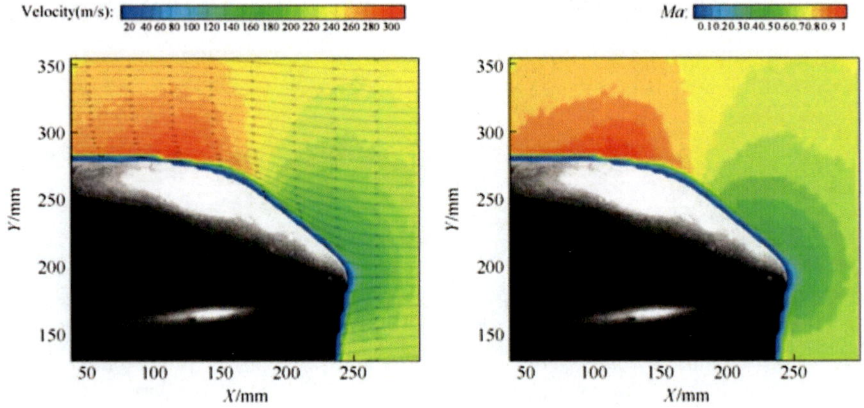

Fig. 7.51 Model-5, $a = 3°$, $Ma = 0.785$ streamline and mach number profile

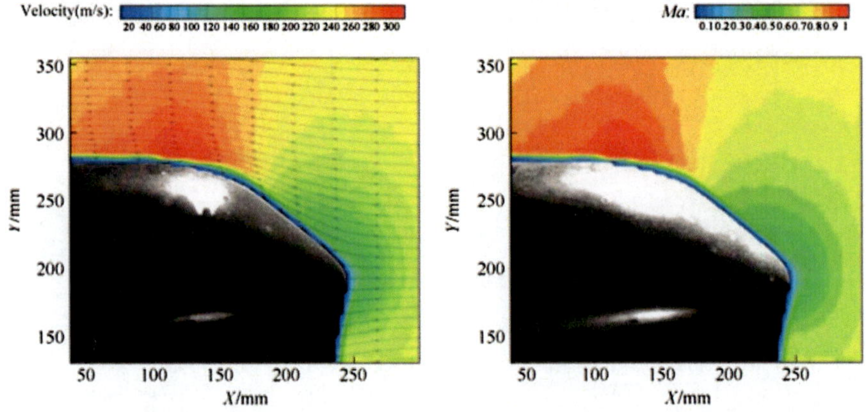

Fig. 7.52 model-5, $a = 3°$, $Ma = 0.820$ streamline and mach number distribution diagram

7# and 8# pressure measuring surfaces when $a = 3°$, $Ma = 0.785$, and the velocity is close to the speed of sound, but does not exceed the speed of sound. The airflow enters the high velocity zone between the 6# and 7# pressure measuring surfaces when $a = 3°$, $Ma = 0.820$, and enters the supersonic velocity zone between the 7# and 8# pressure measuring surfaces, and the supersonic velocity area is very small. The airflow enters the high-velocity area between the 6# and 7# pressure measuring surfaces when $a = 6°$, $Ma = 0.785$, and the velocity is close to the speed of sound, but does not exceed the speed of sound. The airflow enters the high velocity area at the 6# pressure measuring surface, and enters the supersonic velocity area between the 6# and 7# pressure measuring surfaces when $a = 6°$ and $Ma = 0.820$, and the supersonic velocity area is very small.

Comparatively speaking, the position of airflow entering the high velocity region or supersonic region on the model-5 is advanced under the same working conditions,

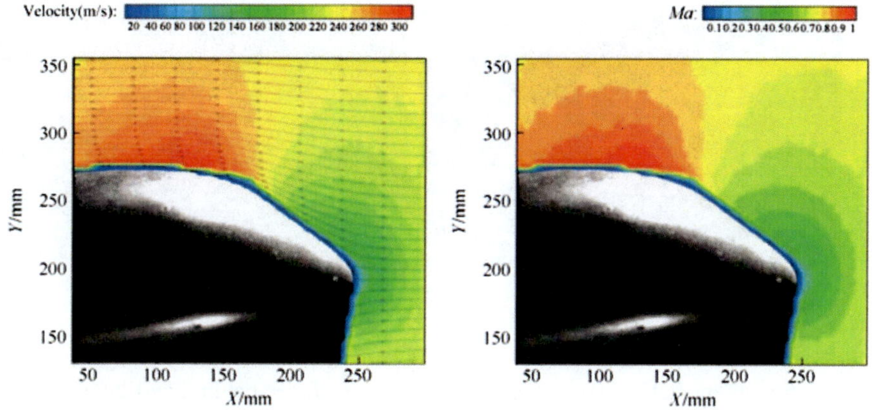

Fig. 7.53 Model-5, $a = 6°$, $Ma = 0.785$ streamline and mach number distribution

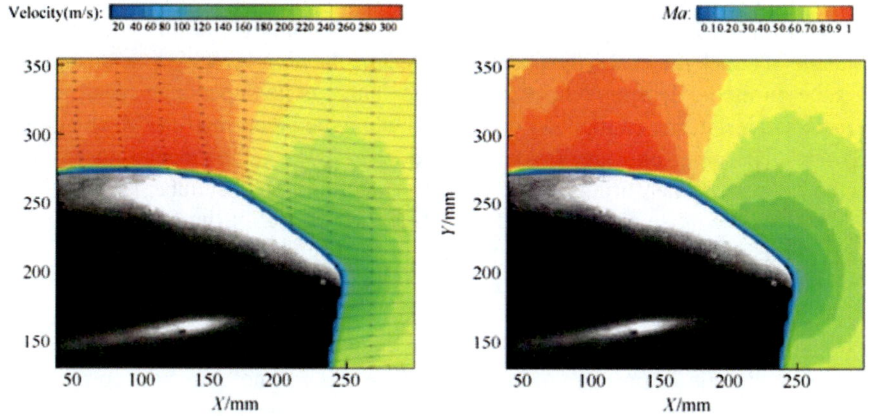

Fig. 7.54 Model-5, $a = 6°$, $Ma = 0.820$ streamline and mach number distribution

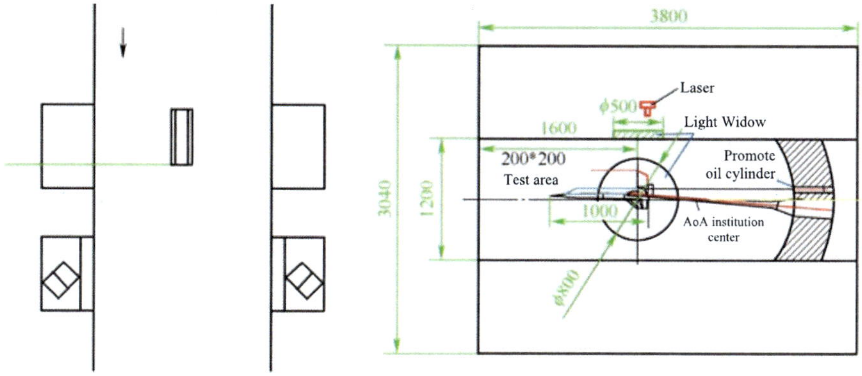

Fig. 7.55 SPIV optical path layout

Fig. 7.56 Laser, light guide arm and sheet light source

which is about half of the pressure measuring surface spacing, and the supersonic region on the model-5 is larger than that on the model-1. The forward movement of the position in the high-velocity zone or supersonic zone on model-5 means that transition and turbulence occur earlier, which leads to increased resistance and energy consumption during flight. In the case of the same incoming flow Mach number, as the angle of attack increases, the position that enters the high velocity zone or supersonic zone on model-1 or model-5 will advance. There is no supersonic speed zone and no wave resistance on model-1 or model-5 when the incoming flow Mach number is 0.785. The supersonic area (maximum $M\alpha_{max} = 1.04$) on the model-1 nose is behind the 7# pressure measuring surface When the incoming Mach number is 0.820, and the aerodynamic noise produced will not affect the comfort of the cockpit (the cockpit porthole is in 3# and 7# between pressure measuring surface).

The airflow enters the supersonic velocity zone between the 6# and 7# pressure measuring surfaces on model-5 (maximum $M\alpha_{max} = 1.05$) when $a = 6°$, $Ma = 0.820$, and enters the sub-surface between 9# and 10# pressure measuring surfaces. It is behind the 7# pressure measuring surface and will not affect the comfort of the cockpit in the sound velocity zone, even if there is a shock wave, it is a weak shock wave.

From the measurement results of velocity field distribution, the airflow gradually accelerates upward along the symmetrical face of the leading edge of the nose, and the velocity distribution is smooth and there is no sudden change, which indicates that there is no flow separation in the flow field near the nose. This is consistent with the results of the pressure distribution and oil flow display experiments performed simultaneously.

The final experiment of the distribution field has obtained the rate field distribution results of the two nose models under multiple working conditions, and obtained the law of speed with the angle of attack and Mach number. Only under the fast cruising Mach number, there is a small area of sound velocity at the top of the nose, with very low intensity, far from the rear of the nose porthole, and there is no flow separation phenomenon. Judging from the measurement results, the model-l nose is better than the model-5 nose. The test data conforms to the aerodynamic law. The DPIV measurement error of this test is 8–0.05 pixels (equivalent to rate error W2.4 m/s), the absolute versus value of the wind tunnel angle of attack mechanism error is: $\Delta aW3'$, test data It can be used as the basis for the pneumatic design of the nose.

7.3.3.3 Application of DPIV/SPIV Technology in Internal Resistance Measurement of Ventilation Model

The engine and airframe of the hypersonic aircraft must be integrated in order to obtain high lift-to-drag aerodynamic performance. The lower surface of the front of the fuselage is also the front compression surface of the super-combustion engine inlet, and the lower surface of the rear of the fuselage also serves as a part of the engine tail nozzle. A large number of wind tunnel tests are needed for the aircraft in order to provide aerodynamic characteristic data of aircraft for control and trajectory. Traditionally, for aircrafts with air-breathing engines, this work is completed through ventilation model force measurement tests, but for hypersonic aircrafts, the traditional test methods have new challenges.

Ventilation model test is an important means to study the influence of engine intake on the aerodynamic characteristics of aircraft. The external flow influence of the model and the momentum change after the air flow passes through the intake port can be determined by simulating the flow in front of the intake port as much as possible. The resistance measured by the balance includes the internal resistance coefficient Cm of the pipeline in the ventilation model test, which must be deducted. The internal resistance coefficient Cam of the model is calculated by measuring the average static pressure P_e and the average total pressure $\overline{P}_{e0}$ at the inner flow outlet. The front resistance of the aircraft matches the net thrust provided by the engine manufacturer in the definition of resistance in this ventilation model test, and the force test data can be directly applied.

The difficulty of hypersonic vehicle ventilation model test technology mainly lies in the difficulty of measuring the airflow parameters at the outlet of the inlet flow. This is due to the expansion shape of the engine tail nozzle, which makes the outlet flow direction of the internal flow not consistent to one direction, while the existence of the balance strut leads to more complicated outlet flow. The most direct effect of inaccurate measurement of airflow parameters at the internal flow outlet is that the internal resistance cannot be reasonably deducted, and the further effect is that the incomplete simulation of internal flow has an effect on the aerodynamic coefficient and moment coefficient of the whole bomb. There are many forms of inlet/engine configurations for hypersonic aircrafts. It is an urgent requirement for model design to study internal resistance deduction technology for different configurations.

The usual internal resistance measurement method is to measure the total static pressure of the internal flow at the nozzle outlet of the engine, calculate the average value of total pressure $\overline{P}_o$ and the average value $\overline{P}$ of static pressure, and use the momentum theorem to calculate the internal resistance of the engine. The relationship between flow velocity, Mach number, total pressure, and static pressure under ideal gas adiabatic conditions can be used to measure the flow field and total pressure field at the outlet of the hypersonic aircraft engine nozzle through SPIV and the total pressure gauge rack to derive the Mach number distribution And static pressure distribution, finally calculate the internal resistance of the engine.

In the test of measuring internal resistance using SPIV technology, the engine outlet velocity field and total pressure are to be measured. *Poi* (*I* Is the total pressure measuring point number), the velocity field is obtained by PIV test, and the total pressure is measured by the pressure rake. The distribution of the total pressure pipe should be as dense as possible. The scalar rate work at any total pressure measurement point i can be obtained on the rate field measured by PIV.

According to the definition of total temperature, there are:

$$T_0 = T_i + \frac{v_i^2}{2c_p} \tag{7.29}$$

In this equation:

T_0 total air temperature (K);
T_i Static temperature at the *i*-th measuring point (K);

Scalar rate at the *i*-th measuring point of T_i (m/s).
Again

$$c_p = \frac{\gamma}{\gamma - 1}\mathrm{R}. \tag{7.30}$$

In this equation:

γ The specific heat ratio of gas is 1.4 for complete gas.

It can be checked in the thermodynamic parameter table of air according to the actual gas state range if the real gas effect is considered. Generally speaking, it does not change much within a certain static temperature range and can be considered as a constant.

R General gas constant (J/Kg).

According to the calculation equation of sound velocity, that is:

$$c = \sqrt{\gamma \mathrm{RT}} \tag{7.31}$$

Solve T in Eq. (7.29) and Eq. (7.30) and substitute Eq. (7.31) to obtain the sound velocity at the i measuring point:

$$c_i = \sqrt{\gamma R\left(T_0 - \frac{v_i^2}{2c_p}\right)} = \sqrt{\gamma \mathrm{R}\left(T_0 - \frac{(\gamma - 1)v_i^2}{2\gamma R}\right)} \tag{7.32}$$

The Mach number of i measuring point can be obtained from the above equation:

$$Ma_i = \frac{v_i}{c_i} = \frac{v_i}{\sqrt{\gamma R\left(T_0 - \frac{(\gamma-1)v_i^2}{2\gamma R}\right)}} \tag{7.33}$$

The gas total static pressure parameter ratio has the equation:

$$\frac{p_0}{p} = (1 + \frac{\gamma - 1}{2}Ma^2)^{\frac{\gamma}{\gamma-1}} \tag{7.34}$$

P0 needs to be measured by the total pressure rake; in general, γ is just 1.4, and the equation is further derived with $y = 1.4$. When $Ma_i \leq 1$, the static pressure of the first/measurement point P_i can be obtained:

$$p_i = p_{0i}\left(1 + \frac{\gamma - 1}{2}Ma_i^2\right)^{-\left(\frac{\gamma}{\gamma-1}\right)} = p_{0i}\left(1 + 0.2Ma_i^2\right)^{-3.5} \tag{7.35}$$

When $Ma_i > 1$, since the measured total pressure p_{0i} is the total pressure after the shock wave, after the total pressure after wave and front wave is converted, the static pressure P_i is calculated as follows:

$$p_i = p_{0i}(\frac{5}{6Ma_i^2})^{3.5}(\frac{7Ma_i^2 - 1}{6})^{2.5} \tag{7.36}$$

So far, according to Eqs. (7.33), (7.35), (7.36), the total pressure measurement points position can be obtained.

Using the value of these measuring points, the average Mach number Ma_e and static pressure p_e of the outlet can be obtained by the average method or weighted average according to the measuring point distribution. The internal resistance can be calculated by the following equation:

$$C_{Ain} = \frac{X_{in}}{q_\infty S_r} \tag{7.37}$$

$$X_{in} = 1.4MaMa_eA_e\overline{P_e}\left[\frac{(1 + 0.2Ma_e^2)^{1/2}}{(1 + 0.2Ma^2)^{1/2}} - \frac{Ma_e}{Ma}\right] - \left(\overline{P_e} - P_\infty\right)A_e \tag{7.38}$$

In this equation:

C_{Ain} internal resistance coefficient;
X_{in} internal resistance (N);
q_∞ free flow pressure (Pa);
S_r reference area (m^2);
A_e cross-sectional area at the outlet of the internal flow (m^2);
Ma free flow Mach number;
Ma_e Mach number at the inflow exit;
P_e free flow static pressure (Pa).

Compared with the traditional internal resistance measurement method, the PIV internal resistance measurement method has the following advantages and innovations:

Using PIV technology, even if there are struts, the velocity distribution of complex flow field at the outlet can be measured more accurately. There is no need to measure static pressure, as long as the total pressure is measured, the internal resistance can be calculated, and the total pressure measurement error is small, which is conducive to improving the accuracy; It is of exploration value to try to apply PIV test technology to engineering practice.

(1) FD-12 wind tunnel ventilation model outlet flow field SPIV measurement test conditions and layout

In the SPIV measurement experiment of ventilation model internal resistance, two cameras are used to simultaneously photograph the illuminated area in the flow field from two off-axis directions (satisfying Scheimpflug condition), and the particle image is recorded to obtain stereo three-dimensional velocity field information. Two CCD cameras (equipped with stereo lens angle control mechanism, and the lens is telephoto lens) are set off-axis at the observation windows on both sides of FD-12 wind tunnel in order to obtain the three-dimensional velocity field of the outlet section of the tail nozzle. Laser light is incident from the observation window on one side of FD-12 wind tunnel to illuminate the outlet section of the tail nozzle. The included angle between the two CCD cameras and the outlet section of the tail nozzle should be 90–120° (Fig. 7.55).

There are two ways to transmit the laser sheet light: ① The laser sheet light SlavedFD-12 enters the observation window on the side of the wind tunnel to illuminate the exit section of the tail nozzle. This is the simplest, but there may be forward and backward scattering problems; ② The laser is installed in the chamber of the special test section of the FD-12 wind tunnel through the vibration reduction table. The laser light is projected to the exit section of the model tail nozzle through the observation window on the upper wall of the test section, and the model is installed reversely.

In SPIV test, all optical equipment should be damped. CCD camera should be installed on a multi-degree-of-freedom mobile platform considering the adjustment of positive and negative power angles in the experimental model, and a multi-degree-of-freedom correction plate should also be manufactured.

The photo of the experimental site is shown in Fig. 7.56.

According to the above-mentioned basic requirements for supersonic SPIV experiment and SPIV measurement scheme for internal resistance of supersonic vehicle ventilation model, in FD-12 wind tunnel, aiming at $Ma = 4$, angle of attack $-4° \leq \alpha \leq 4°$ to carry out research on internal resistance measurement technology of ventilation model. The SPIV incoming flow test conditions are shown in Table 7.9.

(2) Experimental Results and Analysis

In the experiment, the translation calibration plate method is used to generate a spatial stereo grid array of calibration points that meets the requirements of quantity and spatial distribution. The origin of the three-dimensional object surface space coordinate system is established on the object surface of the lens, and its X-axis and Y-axis are the horizontal direction and vertical direction of the object surface and the correction plate plane respectively, the Z-axis is the normal direction of the object surface, and the positive direction of the three-axis conforms to the right-hand rule. It is defined as zero position ($z = 0.0$ mm) when the calibration plate is on the object surface of the lens, and the row and column directions of the 9×9 lattice are coincident with the x-axis and y-axis of the object plane coordinate system respectively.

The calibration plate is shifted four times by a distance of 0.5 mm along the positive direction of the z-axis from the position of $z = -1.0$ mm, thus establishing a three-dimensional grid array of calibration points with a size of 120 mm $(x) \times 120$ mm $(y) \times 2$ mm (z) and a number of calibration points of $9\ (x) \times 9\ (y) \times 5\ (z)$. At the same time of the calibration plate translation, the two cameras on the left and right sides also collected the calibration plate images at the following five z-direction positions, namely-1.0 mm, −0.5 mm, 0.0 mm, 0.5 mm and 1.0 mm, and obtained five groups of images that can reflect the z-direction position changes of the calibration plate lattice. The zero positions calibration plate image is called the reference calibration plate image (Fig. 7.57), because the position calibration plate image will be used to calculate the geometric correction coefficient table of the distorted image. The correction coefficient table uses the two-dimensional bi-harmonic sample based on Green's function. The interpolation method can be obtained. Then the correction coefficient table and bicubic interpolation algorithm are applied to correct the captured images at zero position and other four positions. A calibration plate image of the same size as the standard calibration plate image can be obtained. However,

Table 7.9 Incoming flow test parameters

Ma	P_0/MPa	T_0/K	P_∞/kPa	q_∞/kPa	Re/L/1/m
4.00	6.47	300	4.26	47.73	3.03×10^7

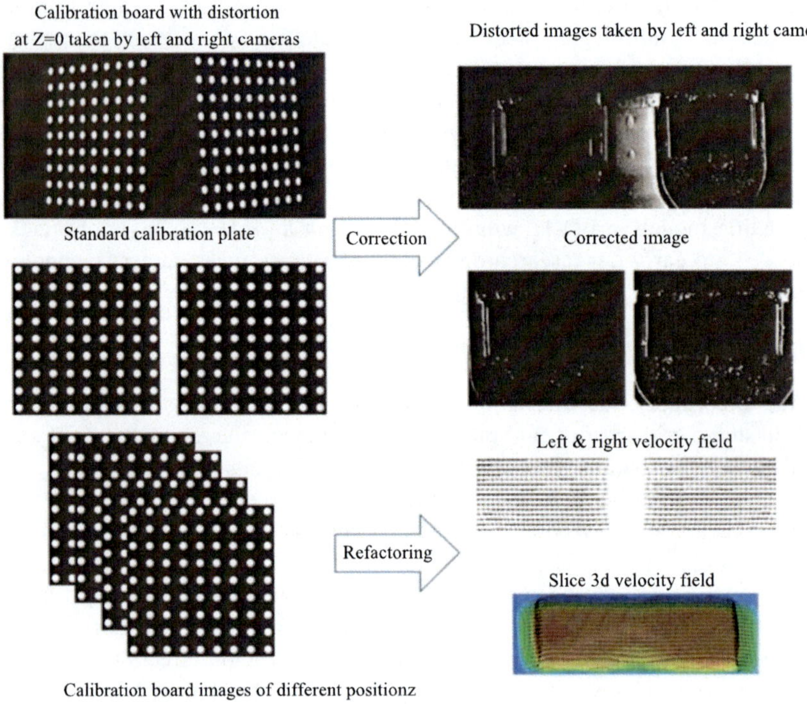

Fig. 7.57 Basic data transactions process

the position has obviously shifted, and the z-direction displacement information of the actual calibration plate is included in the offset of the corrected image. Using these offsets and the z-direction displacement of the object space, various coefficients in the aforementioned calibration algorithm can be obtained, which can be used in the process of forming the two-dimensional velocity field into the three-dimensional velocity field.

The nozzle outlet cross-section coordinate system is defined as follows: based on the geometric center of the outlet cross-section point O is the origin of the coordinate, OX The right wall surface of the axis pointing to the outlet is positive, the OY axis pointing to the upper wall surface of the outlet is positive, OZ axis points to the outside of the outlet, as shown in Fig. 7.58 (rear view of the model).

Figure 7.59 shows that the test is under way. $Ma = 4, a = 0°$, A pair of tracer particle images at intervals of 600 ns. Using the transactions software based on the principle of related algorithms to perform transactions, the rate vector distribution map of the tail nozzle outlet can be obtained (Fig. 7.60).

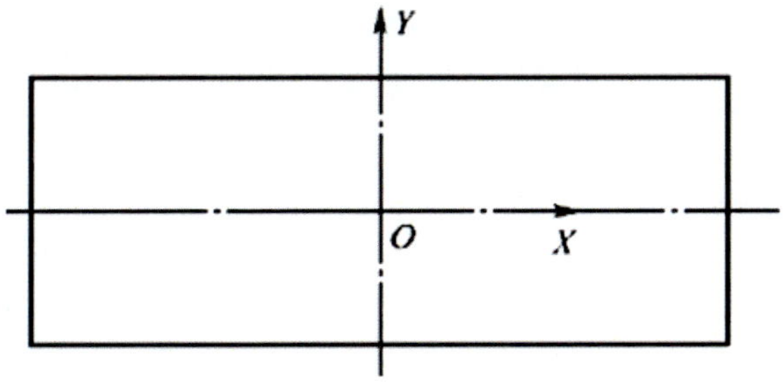

Fig. 7.58 Outlet section coordinate system

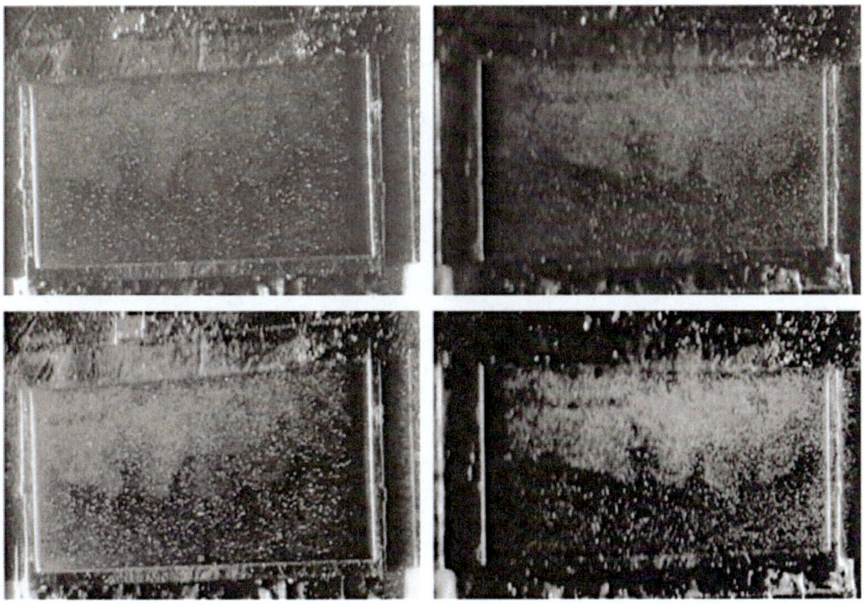

Fig. 7.59 A pair of tail nozzle outlet tracer particle images of left and right cameras using SPIV technology

Figure 7.61 shows *Ma* = *4,* the Mach number cloud images of the model outlet section at different angles of attack show that the Mach number cloud images measured by SPIV have good left-right symmetry. The flow velocity near the wall is significantly lower than that in the central region due to the influence of boundary layer, showing a trend that the central velocity is higher than that on both sides. As the angle of attack changes from negative to positive, the inlet capture flow increases, and the area of the nozzle outlet high-speed region increases, with an angle of attack of 4. The total pressure loss of the flow is relatively large when passing through the inlet, and the area of the low-velocity area on the upper wall increases; at $-4°$ angle of attack, the actual capture area at the inlet of the inlet is significantly reduced, and

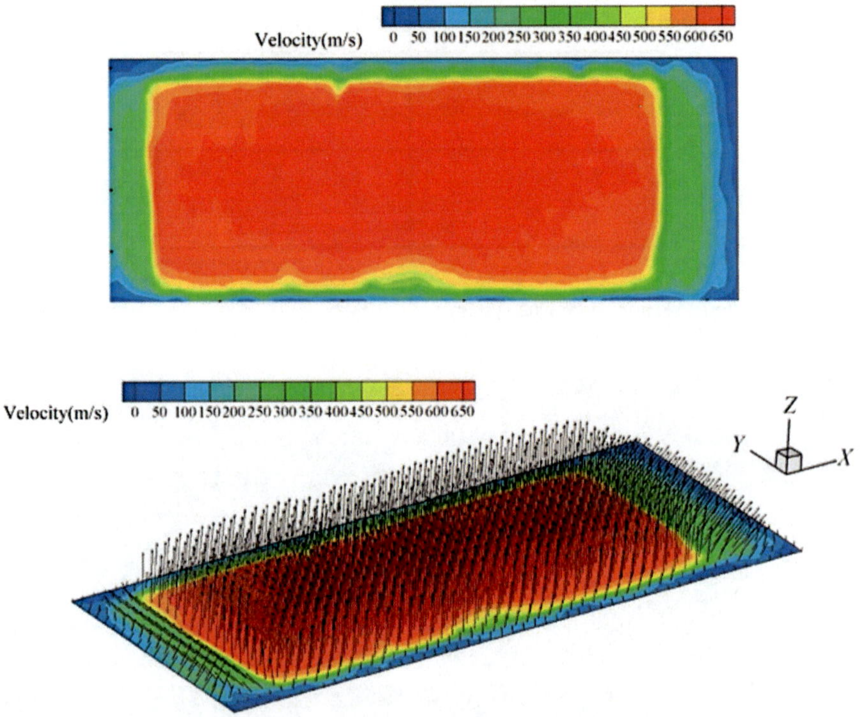

Fig. 7.60 $Ma = 4$, 0° velocity distribution nephogram and three-dimensional velocity vector diagram at the outlet of a nozzle at an angle of attack

the low-velocity area of the nozzle outlet increases. It can be seen that the flow area rate at the exit Master of the model is basically along the normal direction of the exit section.

7.3.4 Summary and Prospect of Particle Image Velocimetry Technology

Particle image velocimetry can obtain the flow structure and velocity distribution of the external turbulent flow field of complex aircrafts due to its advantages of non-contact and global measurement. It is suitable for a wide speed range (low speed, sub-transonic supersonic and hypersonic wind tunnels), and the absolute speed of the measured flow field can reach up to 1500 m/s. It has the ability to display and measure the external flow field for wind tunnel tests of various hypersonic aircrafts.

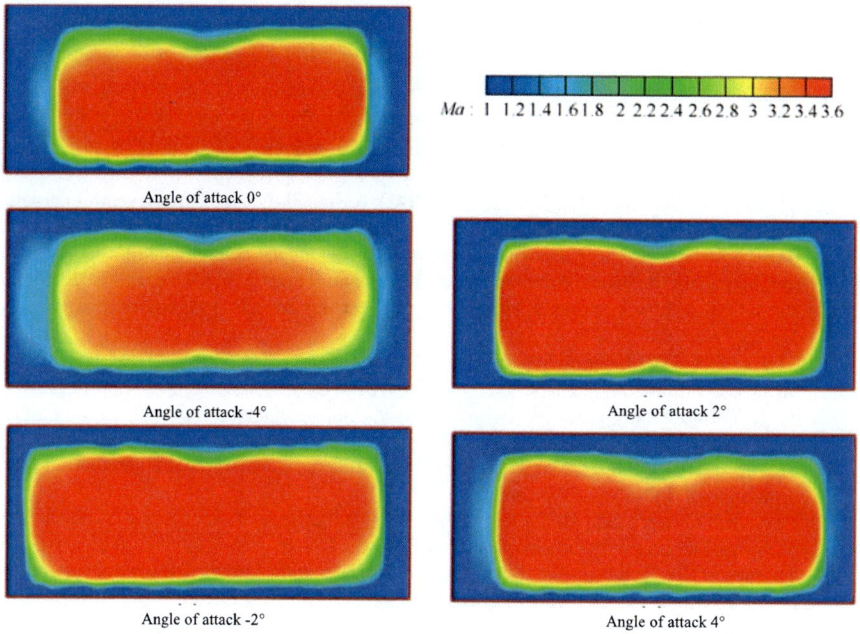

Fig. 7.61 Mach number distribution cloud picture of nozzle outlet section

Its complex space flow field puts forward higher requirements for particle image velocimetry technology with the increasing demand for the development of domestic space models. At present, particle image velocimetry technology is developing towards the direction of three-dimensional volume space velocity field measurement and ultra-high frequency measurement. For example, tomographic particle image velocimetry (TOMOPIV) illuminates the flow field through a volume light source and uses three or four CCD cameras to photograph the flow field from different directions, thus obtaining the distribution of the three-dimensional velocity field in the three-dimensional volume space. MHz Particle Image Velocimetry (MHz-PIV) realizes the acquisition of continuous multi-tracer particle images of flow field with very short time intervals (<1 s) through multi-cavity synchronous high-frequency pulse lasers, ultra-high-speed multi-cameras and high-precision synchronous controllers, making it possible to acquire time-resolution velocity fields of supersonic and even hypersonic flow fields.

References

1. Garlick G (1949) Luminescent materials. Clarendon, Oxford, England, UK 1(6):18–24
2. Czysz PA, Dixon WP (1968) Instrum Control Syst
3. Borella HM (1986) Technical report no. M-2657, March 14, 1986. EG&G Energy measurements Inc.
4. Allison SW, Abraham MA, Boatner LA (1992) Technical digest of the eleventh international congress on lasers & electro-optics laser institute of America. Anaheim, CA
5. Berry SA, Bouslog SA, Brauckmann GJ (1997) Boundary layer transition due to isolated roughness: shuttle results from the larc 20-inch mach 6 tunnel. AIAA Paper 97-0273 1:12–16
6. Buck GM (1989) Automated thermal mapping techniques using chromatic image analysis. NASA TM - 101554
7. Miller CG (1990) Langley hypersonic aerodynamic/aerothermodynamic testing capabilities: present and future. AIAA paper no. 90-1376 3:56–60
8. Scott A, Berry RJ et al (2004) Boundary layer control for hypersonic airbreathing vehicles. AIAA paper 2004-2246 3:34–36
9. Thompson RA (2000) Review of X-33 hypersonic aerodynamic and aerothermodynamic development. In: ICAS 2000 congress
10. Horvath TJ, Berry SA, Hollis BR (1999) X-33 experimental aeroheating at mach 6 using phosphor thermography. AIAA paper 99-33908 23:15–18
11. Thompson RA, Hamilton HH, Berry SA (1998) Hypersonic boundary-layer transition for x-33 phase II vehicle. AIAA P 98-0867 6:12–15
12. Horvath TJ, Scott A et al (2006) Shuttle damage/repair from the perspective of hypersonic boundary layer transition-experimental results. AIAA 2006-2919 5:6–12
13. Michelle LJ, Scott AB (2013) Thermographic phosphor measurements of shock-shock interactions on a swept cylinder. In: Proceedings of TFAWS
14. Dennis CB (2012) Boundary-layer instability measurements in a Mach-6 quiet tunnel. AIAA 2012-3147 6:45–48
15. Derek SL (2008) Aerothermodynamic testing of protuberances and penetrations on the NASA crew exploration vehicle heat shield. AIAA 2008-1240 1:26–27
16. Inna K, Joseph D (2011) Temperature—sensitive paint calibration methodology developed at AEDC tunnel 9. AIAA 2011-851
17. Bi Z, Han S, Wu C et al (2013) Research on phosphorescence thermogram heat measurement technology. Exp Fluid Mech 27(3):87–92
18. Bi Z, Gong J, Wen S et al (2013) Application of phosphor thermography technology in the surface heat flow measurement experiment of imitating X-33 aircraft. In: Proceedings of China mechanics congress
19. Han SG, Wen S, Wu CH et al. Global heat-flux measurements using phosphor thermography technique in gun tunnel. AIAA 2015-3517
20. Fan J et al (1997) Flow display and measurement. Mechanical Industry Press, Beijing
21. Dudderartd MR, Simpkins PG (1986) Laser speckle velocimetry. In: U.S. National Congress of applied mechanics, 10th, Austin, TX, 16–20 June 1986. Proceedings (A87-40051 1731). American Society of Mechanical Engineers, New York (1987), pp 187–200
22. Adrian RJ (1984) Scattering particle characteristics and their effect on pulsed laser measurements of fluid flow Speckle velocimetry vs particle image velocimetry. Appl Opt 23:1690–1691
23. Jacquot P, Rastogi P (1981) Influence of out-of-plane deformation and it's elimination in white light speckle photography. Opt Lasers Eng 2:33–55
24. Sinha S (1988) Improving the accuracy and resolution of particle image or laser speckle velocimetry. Exp Fluids 6:67–68
25. Shen GX, Kang Q (1995) Stereoscopic particle image velocimetry technique and preliminary application for 3d vortex flow. Flow visualization VII. In: Proceedings of the 7th international symposium on flow visualization, Seattle, WA, United States, 11–14 Sept 1995, pp 733–738
26. Hinsch KD (1995) Three-dimensional particle velocimetry. Meas Sci Technol 6:742–753

27. Prasad AK, Jensen K (1995) Scheimpflug stereocamera for particle image velocimetry in liquid flows. Appl Opt 34:7092–7099
28. Westerweel J, Van Oord J (1999) Stereoscopic PIV measurements in a turbulent boundary layer. In: Stainslaus M, Kompenhans J, Westerweel J (eds) Particle image velocimetry: progress toward industrial application. Kluwer, Dordrecht
29. Shaobo Peng (1998) Development and preliminary debugging of a pneumatic particle generator. Beijing University of Aeronautics and Astronautics, Beijing
30. Dennis R (1988) Aerosol handbook. Atomic Energy Press, Beijing
31. Van HC (1957) Light scattering by small particles. Dover Publications, New York
32. Raffel M, Willert CE, Kompenhans J (1998) Particle image velocimetry: a practical guide. Springer, Berlin, New York
33. Humphreys WM, Bartram SM, Blackshire JL (1993) A survey of particle image velocimetry applications in Langley aerospace facilities. AIAA, Aerospace Sciences Meeting and Exhibit, 31st, Reno, NV, pp 11–14
34. Lei YF (2010) Ultra-fine jet pulverization technology. Mater Sci Eng 18(4):145–149
35. Keane RD, Adrian RJ (1988) Double exposure, multiple-field particle image velocimetry for turbulent probability density. Opt Lasers Eng 9:211–228
36. Yao CS, Adrian RJ (1984) Orthogonal compression and 1-D analysis technique for measurement of 2-D particle displacements in pulsed laser velocimetry. Appl Opt 23:1687–1689
37. Ronald JA (1986) Image shifting technique to resolve directional ambiguity in double-pulsed velocimetry. Appl Opt 25:3855–3858
38. Keane RD, Adrian RJ (1990) Optimization of particle image velocimetry. I. Double pulsed systems. Meas Sci Technol
39. Liu ZC, Landreth CC, Adrian RJ et al (1991) High resolution measurement of turbulent structure in a channel with particle image velocimetry. Exp Fluids 10(6):56–59
40. Willert CE, Gharib M (1991) Digital particle image velocimetry. Exp Fluids 10(4):181–193
41. Westerweel J (1993) Digital particle image velocimetry: theory and application. Dissertation Mechanical Maritime and Materials Engineering
42. Westerweel J (1997) Fundamentals of digital particle image velocimetry. Meas Sci Technol 8:1379
43. Westerweel J (2000) Theoretical analysis of the measurement precision in particle image velocimetry. Exp Fluids 29(Supplement 1):45–47
44. Chen J, Katz J (2005) Elimination of peak-locking error in PIV analysis using the correlation mapping method. Meas Sci Technol 16:1605–1618
45. Adrian RJ (2005) Twenty years of particle image velocimetry. Exp Fluids 39(2):159–169